Handbook for Inspection of Ships and Issuance of Ship Sanitation Certificates

Handbook for Inspection of Ships and Issuance of Ship Sanitation Certificates

WHO Library Cataloguing-in-Publication Data
International health regulations (2005): handbook for inspection of ships and issuance of ship sanitation certificates.

1. Legislation, Health. 2. Communicable disease control - legislation. 3. Ships. 4.Sanitation. 5. Disease outbreaks - legislation. 6. Disease transmission - prevention and control. 7. Handbooks. I. World Health Organization.

ISBN 978 92 4 154819 9 (NLM classification: WA 810)

Printed in Italy
WHO/HSE/IHR/LYO/2011.3

Design by Crayonbleu, Lyon, France
Editing by Biotext Pty Ltd, Canberra, Australia

CONTENTS

FOREWORD

On 23 May 2005, the Fifty-eighth World Health Assembly adopted the International Health Regulations (IHR) (2005), and the Deratting Certificate/Deratting Exemption Certificate required by the IHR (1969) was replaced by the broader-scope ship sanitation certificates (SSCs), which came into force on 15 June 2007.

The IHR (2005) states that parties can authorize certain ports to issue the SSCs and their extensions, as well as to provide the services referred to in Annex 1 of the regulations. The authorized ports should have, among other capacities, trained personnel available to board a ship and to identify any significant risk to public health, as well as to take control measures. Thus, it is imperative to have a global, standardized operational procedure for inspecting ships.

After the IHR (2005) took effect in June 2007, the World Health Organization (WHO), Health Security and Environment (IHR ports, airports and ground crossings) developed the *Interim technical advice for inspection and issuance of ship sanitation certificates*. This technical advice, which was published in August 2007, assists States Parties to manage ship inspection and issue SSCs.

This document, *Handbook for inspection of ships and issuance of ship sanitation certificates* (the handbook), replaces the previous interim technical advice, and reflects the need for a common understanding of the purpose and scope of the application of SSCs worldwide. It is an important tool for helping to prevent and control known public health risks (not just rodents), and provides a common way to register and communicate events and measures taken on board. The handbook is intended to raise conveyance operators' awareness and response to public health events, and provide the opportunity for routine verification of health status on board at least twice each year.

This handbook may be used in conjunction with the *Guide to ship sanitation* (WHO, 2011) and the International medical guide for ships (WHO, 2007), which are oriented towards preventive health and curative health, respectively, on board ships.
The handbook was developed through an iterative series of drafting and peer-review steps. The following expert meetings were held to revise the handbook:

- informal Transportation Working Group Meeting for Ship Sanitation Certificates, Lyon, France, 6–8 November 2007;
- informal Transportation Draft Working Group Meeting on procedures for inspection and issuance of ship sanitation certificates, Lyon, France, 17–19 December 2007;
- informal consultation for draft on procedures for inspection and issuance of ship sanitation certificates, Lyon, France, 14–16 April 2008;
- meeting on recommended procedures for inspection and issuance of ship sanitation certificates, Lyon, France, 14–15 April 2009;
- informal consultation meetings for the ship sanitation guidelines, Lyon, France, 12–16 October 2009.

A public consultation version of this handbook was posted on the WHO website in May 2010. During the course of meetings and peer review, participants and experts from represented cruise ship operators, seafarer associations, collaborating Member States for the IHR (2005), Port State Control, port health authorities and other regulatory agencies were involved from diverse developing and developed countries. The acknowledgements section contains a complete list of contributors.

From 2008 to 2010, several workshops and field activities were held at the subregional, regional and interregional level. Experts from all WHO regions participated, and the workshops provided an opportunity to revise previous interim technical advice and test the new draft handbook using training exercises onboard ships. The workshops and field activities were supported by WHO regional and country offices, as well as public health authorities in different countries, including Sines, Portugal (2009); Santos (2008), Fortaleza (2010), Brazil; Palma de Majorca (2008), Cartagena (2009), Las Palmas de Gran Canaria (2010), Spain; Amsterdam, the Netherlands (2007); Hamburg, Germany (2008); Miami, United States of America (2008); Bridgetown, Barbados (2008); Manila, the Philippines (2009); Colombo, Sri Lanka (2010); and Paris, France (2009).

ACKNOWLEDGMENTS

This handbook was developed in consultation with experts from diverse developing and developed countries.

The work of the following individuals was crucial to the development of this edition of the Handbook for inspection of ships and issuance of ship sanitation certificate and is gratefully acknowledged:

Jaret T. Ames, Vessel Sanitation Program, Centers for Disease Control and Prevention, Atlanta, United States of America

James Barrow, Division of Global Migration and Quarantine, National Center for Preparedness, Detection, and Control of Infectious Diseases, Centers for Disease Control and Prevention, Atlanta, United States of America

Marie Baville, Environment Health Officer, Department of Emergency Response and Preparedness, General Directorate of Health, Ministry of Health, Paris, France

Priagung Adhi Bawono, Quarantine Sub-directorate, Directorate General Disease Control and Environmental Health, Ministry of Health, Jakarta, Indonesia

David Bennitz, Public Health Bureau, Health Canada, Ottawa, Canada

Colin Browne, Pan American Health Organization/Eastern Caribbean Countries, Bridgetown, Barbados, World Health Organization

Luiz Alves Campos, National Health Surveillance Agency (Anvisa), Brasilia, Brazil

Susan Courage, Environmental Health Bureau, Safe Environments Directorate
Health Canada, Canada

Yves Chartier, WHO, Geneva, Switzerland

Frédéric Douchin, Departmental Directorate of Health and Social Affairs of the Seine Maritime, France

Zhiqiang Fang, Department of Health Quarantine of General Administration of Quality Supervision, Inspection and Quarantine, Beijing, China

Milhar Fuazudeen, Maritime Training and Human Element Section, Maritime Safety Division, International Maritime Organization, London, United Kingdom

Christos Hadjichristodoulou, University of Thessaly, Larissa, Greece

Daniel Lins Menucci, WHO, Lyon, France

Hameed Gh H Mohammad, Ports and Borders Health Division, Rumaithiya, State of Kuwait

Rosemarie Neipp, General Directorate for Public Health and Foreign Health Affairs, Ministry of Health and Social Policy, Spain

Ma Lixin, Department of Health Quarantine of General Administration of Quality Supervision, Inspection and Quarantine, Beijing, China

Henry Kong, Port Health Office, Hong Kong Special Administrative Region, China

Jenny Kremastinou, National School of Public Health, Athens, Greece

Maike Lamshöft, Hamburg Port Health Center, Central Institute for Occupational Medicine and Maritime Medicine, Hamburg, Germany

Fábio Miranda da Rocha, National Health Surveillance Agency (Anvisa), Brasilia, Brazil

Mohamed Moussif, Mohamed V International Airport, Casablanca, Morocco

Barbara Mouchtouri, University of Thessaly, Larissa, Greece

Matthijs Plemp, National Institute of Public Health and the Environment, The Netherlands

Thierry Paux, Department of Alert, Response and Preparedness, Ministry of Health, Paris, France

Tobias Riemer, Hamburg Port Health Center, Central Institute for Occupational Medicine and Maritime Medicine, Germany

Clara Schlaich, Hamburg Port Health Center, Central Institute for Occupational Medicine and Maritime Medicine, Hamburg, Germany

Christoph Sevenich, Institute for Occupational and Maritime Medicine, Hamburg Port Health Center, Germany

Natalie Shaw, International Shipping Federation, London, United Kingdom

Mel Skipp, Carnival UK, Cruise Lines International Association, London, United Kingdom

Maria Dulce Maia Trindade, Macao International Airport/Port Health Authority, Centre for Prevention and Control of Disease/ Health Bur eau, Government of Macao Special Administrative Region, China

Stéphane Veyrat, Department of Emergency Response and Preparedness, General Directorate of Health, Ministry of Health, Paris, France

Mario Vilar, Ministerio de Salud Publica, Dirección General de la Salud, Montevideo, Uruguay

Ninglan Wang, WHO, Lyon, France

Sandra Westacott, Port Health Services, Southampton City Council, Southampton, United Kingdom

Ruth Anderson, Agnieszka Rivière provided secretarial and administrative support throughout the meetings during the development of the guide. Daniel Lins Menucci, Christos Hadjichristodoulou, Barbara Mouchtouri, Bruce Plotkin, Clara Schlaich, Christoph Sevenich and Ninglan Wang undertook final technical writing and revision roles in developing the guide.The preparation of this document, Handbook for inspection of ships and issuance of ship sanitation certificates, would not have been possible without the generous technical and logistical support of several institutions, including the French Ministry of Health; the Hamburg Port Health Center (Germany); the United States Centers for Disease Control and Prevention,; University of Thessaly, Greece; National Health Surveillance Agency (Anvisa), Brazil; General Administration of Quality Supervision, Inspection and Quarantine(AQSIQ), China; Spain Ministry of Health; Portugal Ministry of Health, and Health Canada.

GLOSSARY

- **Acceptable non-rat-proof material**

A material with a surface that is treated ("flashed") to be resistant to rat gnawing when the edges are exposed to gnawing (the "gnawing-edges"), but that is subject to penetration by rats if the gnawing-edges are not treated.

- **Accessible**

Capable of being exposed for cleaning and inspection with the use of simple tools, such as a screwdriver, pliers or an open-ended wrench.

- **Affected**

Persons, baggage, cargo, containers, conveyances, goods, postal parcels or human remains that are infected or contaminated, or carry sources of infection or contamination, so as to constitute a public health risk.

- **Affected area**

A geographical location specifically for which health measures have been recommended by WHO under International Health Regulations (2005)

- **Airbreak**

A piping arrangement in which a drain from a fixture, appliance or device discharges indirectly into another fixture, receptacle or interceptor at a point below the flood-level rim.

- **Airgap**

The unobstructed vertical distance through the free atmosphere between the lowest opening from any pipe or faucet supplying water to a tank, plumbing fixture or other device and the flood-level rim of the receptacle or receiving fixture. The air gap should typically be at least twice the diameter of the supply pipe or faucet, or at least 25 mm.

- **Backflow**

The flow of water or other liquids, mixtures or substances into the distribution pipes of a potable supply of water from any source or sources other than the potable water supply. Back-siphonage is one form of backflow.

- **Backflow preventer**

A mechanical device installed in a water or waste line to prevent the reversal of flow under conditions of back pressure. An approved backflow-prevention plumbing device is typically used on potable water distribution lines where there is a direct connection or a potential connection between the potable water distribution system and other liquids, mixtures or substances from any source other than the potable water supply. Some devices are designed for use under continuous water pressure, whereas others are non-pressure types. In the check-valve type, the flap should swing into a recess when the line is flowing full to preclude obstructing the flow.

- **Back-siphonage**

The backward flow of used, contaminated or polluted water from a plumbing fixture or vessel or other source into a water-supply pipe as a result of negative pressure in the pipe.

• **Black water**
Waste from toilets, urinals or medical facilities.

• **Biohazard bag**
Bag used to secure biohazard waste that requires microbiological inactivation in an approved manner for final disposal. Must be disposable and impervious to moisture, and have sufficient strength to preclude tearing or bursting under normal conditions of usage and handling.

• **Child care facility**
Facility for child-related activities where children are not yet out of diapers or require supervision using the toilet facilities, and are cared for by vessel staff.

• **Cleaning**
Removal of visible dirt or particles through mechanical action, reducing the microbial environment population through the application of chemical, mechanical or thermal processes for a certain period of time.

• **Closed joints, seams and crevices**
Those where the materials used in fabrication of the equipment join or fit together snugly. Suitable filling materials may be used to effect a proper closure.

• **Communicable disease**
Diseases caused by pathogenic microorganisms, such as bacteria, viruses, parasites or fungi. The diseases can be spread, directly or indirectly, from one person to another. Zoonotic diseases are communicable diseases of animals that can cause disease when transmitted to humans.

• **Competent authority**
Authority responsible for implementing and applying health measures under the International Health Regulations (2005).

• **Corrosion-resistant**
Capable of maintaining original surface characteristics under prolonged influence of the use environment, including the expected food contact and the normal use of cleaning compounds and sanitizing solutions. Corrosion-resistant materials must be non-toxic.

• **Coved**
A concave surface, moulding or other design that eliminates the usual angles of 90 degrees or less.

• **Cross-connection**
Any unprotected actual or potential connection or structural arrangement between a public or a consumer's potable water system and any other source or system through which it is possible to introduce into any part of the potable system any used water, industrial fluid, gas or substance other than the intended potable water with which the system is supplied. Bypass arrangements, jumper connections, removable sections, swivel or change-over devices and other temporary or permanent devices through which, or because of which, backflow can occur are considered to be cross-connections.

- **Disease**

An illness or medical condition, irrespective of origin or sources, that presents or could present significant harm to humans.

- **Disinfection**

The procedure whereby measures are taken to control or kill infectious agents on a human or animal body, on a surface or in or on baggage, cargo, containers, conveyances, goods and postal parcels by direct exposure to chemical or physical agents.

- **Easily cleanable**

Fabricated with a material, finish and design that allows for easy and thorough cleaning with normal cleaning methods and materials.

- **Flashing**

The capping or covering of corners, boundaries and other exposed edges of acceptable non-rat-proof material in rat-proof areas. The flashing strip should typically be of rat-proof material, wide enough to cover the gnawing-edges adequately and firmly fastened.

- **Food contact surfaces**

Surfaces of equipment and utensils with which food normally comes in contact, and surfaces from which food may drain, drip or splash back onto surfaces normally in contact with food. This includes the areas of ice machines over the ice chute to the ice bins.

- **Food display areas**

Any area where food is displayed for consumption by passengers and/or crew.

- **Food handling areas**

Any area where food is stored, processed, prepared or served.

- **Food preparation areas**

Any area where food is processed, cooked or prepared for service.

- **Food service areas**

Any area where food is presented to passengers or crew members (excluding individual cabin service).

- **Food storage areas**

Any area where food or food products are stored.

- **Free pratique**

The permission for a ship to enter a port, embark or disembark, discharge or load cargo or stores.

- **Grey water**

All used water, including drainage from galleys, dishwashers, showers, laundries, and bath and washbasin drains. It does not include black water or bilge water from the machinery spaces.

- **Halogenation**

In this context, halogenation refers to disinfection using halogen disinfectants, such as chlorine, bromine or iodine, to treat recreational water or potable water to reduce the concentration of pathogenic microorganisms.

• International voyage

(a) In the case of a conveyance, a voyage between points of entry in the territories of more than one state, or a voyage between points of entry in the territory or territories of the same state if the conveyance has contacts with the territory of any other state on its voyage, but only as regards those contacts.

(b) In the case of a traveller, a voyage involving entry into the territory of a state other than the territory of the state in which that traveller commences the voyage.

• National IHR Focal Point

The national centre, designated by each State Party, which shall be accessible at all times for communication with WHO IHR Contact Points.

• Non-absorbent materials

Materials with surfaces that are resistant to moisture penetration.

• Non-food contact surfaces

All exposed surfaces, other than food contact or splash contact surfaces, of equipment located in food storage, preparation and service areas.

• Perishable food

A food that is natural or synthetic and that requires temperature control because it is in a form capable of supporting:

• the rapid and progressive growth of infectious or toxigenic microorganisms;

• the growth and toxin production of *Clostridium botulinum*; or, in *raw* shell eggs, the growth of *Salmonella enteritidis*.

• Personal protective equipment (PPE)

Equipment used to create a protective barrier between the worker and the hazards in the workplace.

• Point of entry

A passage for international entry or exit of travellers, baggage, cargo, containers, conveyances, goods and postal parcels, as well as agencies and areas providing services to them on entry or exit.

• Portable

A description of equipment that is readily removable or mounted on casters, gliders or rollers; provided with a mechanical means so that it can be tilted safely for cleaning; or readily movable by one person.

• Potable water

Fresh water that is intended for human consumption, drinking, washing, teeth brushing, bathing or showering; for use in fresh-water recreational water environments; for use in the ship's hospital; for handling, preparing or cooking food; and for cleaning food storage and preparation areas, utensils and equipment. Potable water, as defined by the WHO *Guidelines for drinking-water quality* (2008) does not represent any significant risk to health over a lifetime of consumption, including different sensitivities that may occur between life stages.

- **Potable water tanks**

All tanks in which potable water is stored from bunkering and production for distribution and use as potable water.

- **Public health emergency of international concern**

An extraordinary event that is determined, as provided in the International Health Regulations (2005):
- to constitute a public health risk to other states through the international spread of disease; and
- to potentially require a coordinated international response.

- **Public health risk**

A likelihood of an event that may affect adversely the health of human populations, with an emphasis on one that may spread internationally or may present a serious and direct danger.

- **Readily removable**

Capable of being detached from the main unit without the use of tools.

- **Removable**

Capable of being detached from the main unit with the use of simple tools, such as a screwdriver, pliers or an open-ended wrench.

- **Scupper**

A conduit or collection basin that channels water run-off to a drain.

- **Sealed spaces**

Spaces that have been effectively closed, all joints, seams and crevices having been made impervious to insects, rodents, seepage, infiltration and food fragments or other debris.

- **Seam**

An open juncture between two similar or dissimilar materials. Continuously welded junctures, ground and polished smooth, are not considered seams.

- **Sewage**

According to the internationally accepted definition in the International Maritime Organization's International Convention for the Prevention of Pollution from Ships (MARPOL 73/78), sewage is defined as:
- drainage and other wastes from any form of toilets, urinals and WC scuppers;
- drainage from medical premises (e.g. dispensary, sick bay) via washbasins, wash tubs and scuppers located in such premises;
- drainage from spaces containing living animals (e.g. livestock carriers); or
- other waste waters (e.g. grey water from showers) when mixed with the drainages defined above.

- **Sewage treatment**

The process of removing the contaminants from sewage to produce liquid and solid suitable for discharge to the environment or for reuse. It is a form of waste management. A septic tank or other on-site wastewater treatment system such as biofilters can be used to treat sewage close to where it is created.

The common method of sewage treatment is to flush sewage from toilets through a piping system into a holding tank where it is comminuted, decanted and broken down by naturally occurring bacteria in an aerobic process, and disinfected before it is discharged into the open sea. It is important to consider that an excessive use of cleaners and disinfectant in the sewage system may destroy the natural bacteria in the treatment plant. The aerobic process needs oxygen. Therefore, aerators blow air into the biological compartment. Toxic gases can be produced during the process.

- **Ship**

A seagoing or inland navigation vessel on an international voyage (IHR, 2005).

- **Ship water system**

On-board treatment equipment and facilities, water storage tanks, and all the plumbing and fixtures on the ship.

- **Smooth**

A food contact surface that is free from pits and inclusions, with a cleanability equal to or exceeding that of (100 grit) number 3 stainless steel.
A non-food contact surface that is equal to commercial grade hot-rolled steel and is free from visible scale.
A deck, bulkhead or deck head that has an even or level surface with no roughness or projections that render it difficult to clean.

- **Surveillance**

Systematic ongoing collection, collation and analysis of data for public health purposes, and the timely dissemination of public health information for assessment and public health response as necessary.

- **State Party**

Under the International Health Regulations, "States Parties» are those states that have become bound by the revised International Health Regulations (2005).

- **Traveller**

Natural person taking an international voyage.

- **Turbidity**

Cloudiness or lack of transparency of a solution due to presence of suspended particles. Typically measured in nephelometric turbidity units (NTU).

- **Vector**

An insect or other animal that normally transports an infectious agent that constitutes a public health risk.

- **Verification**

Final monitoring for reassurance that the system as a whole is operating safely. Verification may be undertaken by the supplier, by an independent authority or by a combination of these, depending on the administrative regime of a given country. Typically includes testing for faecal indicator organisms and hazardous chemicals.

ACRONYMS

BWMP	ballast-water management plan
CAC	Codex Alimentarius Commission
FAO	Food and Agriculture Organization of the United Nations
FSP	food safety plan
GDWQ	*Guidelines for drinking-water quality* (World Health Organization)
HACCP	hazard analysis critical control point
HPC	heterotrophic plate count
IHR	International Health Regulations
ILO	International Labour Organization
IMGS	*International medical guide for ships* (International Labour Organization, International Maritime Organization, World Health Organization)
IMO	International Maritime Organization
ISM manual	International Safety Management manual
ISO	International Organization of Standardization
ISPP certificate	International Sewage Pollution Prevention certificate
MEPC	Marine Environment Protection Committee
MFAG	*Medical first aid guide for use in accidents involving dangerous goods* (International Maritime Organization)
PPE	personal protective equipment
SSC	ship sanitation certificate
SSCC	Ship Sanitation Control Certificate
SSCEC	Ship Sanitation Control Exemption Certificate
UV	ultraviolet
WHO	World Health Organization

INTRODUCTION

The International Sanitary Regulations were first adopted in 1951. In 1969, they were renamed as the International Health Regulations (IHR). The 1951 IHR were intended to monitor and control six serious infectious diseases: cholera, plague, yellow fever, smallpox, relapsing fever and typhus. In the intervening 50 years, many developments affected the international transmission of disease, including changes in international ship traffic. Therefore, on 23 May 2005, the World Health Assembly adopted a revised IHR by way of resolution WHA58.3, which entered into force on 15 June 2007.

Beginning with the 1951 IHR, the Deratting Certificate/Deratting Exemption Certificate was a required document for the international public health control of ships visiting international ports. The Deratting Certificate helped to reduce the international spread of rodent-borne diseases, especially plague. All ships on international voyages were required to renew this certificate every six months, and this renewal required all areas of the ship to be inspected. In the IHR (2005), the Deratting Certificate/Deratting Exemption Certificate was replaced by the much broader ship sanitation certificates (SSCs) and was no longer valid after 2007.

The IHR (2005) SSCs are of particular importance for the prevention and control of public health risks on board ships on international voyages. They provide internationally recognized documentation regarding the sanitary conditions of a ship, while reducing the need for further and more frequent inspections of the ship during the period for which the certificate is valid (but with options for additional inspections under certain limited circumstances).

This handbook is intended to be used as reference material for port health officers, regulators, ship operators and other competent authorities in charge of implementing the IHR (2005) at ports and on ships. The handbook is based on the IHR (2005) provisions regarding ship inspection and issue of SSCs. They provide guidance for preparing and performing the inspection, completing the certificates and applying public health measures within the framework of the IHR (2005).[1]

1. IHR Article 2, purpose and scope: "to prevent, protect against, control and provide a public health response to the international spread of disease in ways that are commensurate with and restricted to public health risks, and which avoid unnecessary interference with international traffic and trade".

SCOPE

SSCs are used to identify and record all areas of ship-borne public health risks (not limited to rodents).[2] They require the application of comprehensive and detailed inspection procedures and techniques by personnel who are trained in public health issues. The relevant IHR (2005) provisions include Articles 20, 22, 23, 24, 25, 27, 28, 30, 31, 32, 33, 34, 35, 36, 37, 39, 40, 41, 42, 43, 44 and 45, and Annexes 1, 3, 4 and 5.

SSCs emphasize the criteria to be considered during the inspection of areas of the ship. They also provide information for determining which public health measures should be adopted to prevent and control public health risks on board, in turn preventing the international spread of disease. At the completion of the inspection, a new SSC should be issued—either a Ship Sanitation Control Exemption Certificate or a Ship Sanitation Control Certificate, according to the results of the inspection. If the inspection cannot be accomplished at an authorized port (listed on the World Health Organization [WHO] website), the existing certificate may be extended by no more than one month (this must also be done at an authorized port). As provided in the regulations, when an SSC is issued, there is no distinction as to nationality, flag, registry or ownership of the ship.

Part A of this document is a reference for pre-inspection planning and administrative arrangements for issuing the Ship Sanitation Control Exemption Certificate or the Ship Sanitation Control Certificate. Part A describes activities that are the responsibility of port health officers, and national or local public health authorities. These activities maintain adequate standards for ship inspection and issue of SSCs.

Part B of this document is a template for the inspection and issue of SSCs. It describes the areas to be inspected; the relevant standards that apply; the possible evidence that may be found or sample results that could constitute a public health risk; the documentation that must be reviewed before, during and after the inspection process; and control measures or corrective actions that must be taken. The format of this template follows the IHR model SSC in Annex 3 of the IHR (2005). Part B may also be used as reference material for regulators, ship operators and ship builders, and serve as a checklist for understanding and assessing the potential health impacts of projects involving the design of ships.

Throughout this document, references are provided on occupational health issues for ships' crews or other events that affect public health, and inspections applicable to items of IHR Annex 3. References are also provided on the prevention and control of events that may constitute a public health event of international concern (as defined in the IHR).

This handbook also addresses environmental issues that may constitute public health risks, such as the discharge by ships of sewage, waste and ballast water. Harmful contamination (other than microbial contamination), such as from radionuclear sources, could also be found on ships. Such contamination is subject to the IHR (2005) and the requirements of Article 39 and Annex 3; however, specific control measures are outside the scope of this handbook. This

2. IHR Annex 3 provides a comprehensive list of areas, facilities and systems to be inspected with a view to issue of an SSC.

handbook only address the adoption of preliminary control measures, and the mobilization of specialized experts and agencies to respond to radiological events if detected on board.

In summary, Parts A and B of this document have been developed to assist the competent authority at the port in determining:

- recommended competencies for the personnel inspecting ships for the issue of SSCs;
- administrative arrangements for routine enquiries for the planning of ship inspection and issue of SSCs;
- methods of identification, measurement and control of public health risks related to ships, travellers, cargo or discharge;
- procedures relating to the prevention of international spread of disease;
- recording of information on the SSC, including for further action by the ship's crew or the competent authorities at future ports of call;
- communication and response requirements for events of public health concern, including incidents and emergencies on board.

The Guide to ship sanitation (WHO, 2011) and the *International medical guide for ships* (WHO, 2007) are companion volumes to this document, oriented towards preventive health and curative health, respectively, on board ships.

PART A:
INSPECTION SYSTEM REQUIREMENTS

1. Overview of legal and policy framework

When the International Health Regulations (IHR) (2005) came into force on 15 June 2007, competent authorities could require from international ships the IHR model ship sanitation certificate (SSC) (IHR Annex 3), which covers public health risks on board, and the necessary inspections and control measures taken in accordance with the IHR (2005). Competent authorities are required to use the Annex 3 SSC to identify and record all evidence of contamination or infection and other risks to human health in different areas, facilities or systems, together with any required control measures that must be applied (as authorized by the IHR) to control public health risks[1]. The SSCs may be required from all ships, whether seagoing or inland navigation ships, on international voyages that call at a port of a State Party.

According to the IHR (2005), States Parties authorize certain ports to inspect ships and issue the certificates (or their extensions) and to provide related services and control measures, as referred to in Article 20.3 and Annex 1 of the IHR (2005). Any port authorized to issue the Ship Sanitation Control Certificate (SSCC) must have the capability to inspect ships, issue certificates and implement (or supervise the implementation of) necessary health control measures. States Parties can also authorize ports to issue the Ship Sanitation Control Exemption Certificate (SSCEC) or to grant extensions of up to one month to conveyance operators, if they are unable to carry out the necessary measures at the port in question.

The IHR (2005) requires States Parties to ensure that all SSCs are issued according to IHR Article 39 and Annex 3.
The States Parties must also send to the World Health Organization (WHO) the list of their ports authorized to:

- issue SSCCs and provide the related services referred to in IHR (2005) Annex 3 (Requirements for the SSC) and Annex 1B (Core capacity requirements for designated ports);
- issue SSCECs only, and extend a valid SSCEC or SSCC for one month until the ship arrives in a port at which the certificate may be issued.

Each State Party must inform WHO of any changes that occur in the status of the listed ports. WHO publishes and updates a list of these authorized ports, with related information. This list is available on the WHO (IHR 2005) website (http://www.who.int/ihr/ports_airports/en).

According to the IHR (2005), the SSCs of each State Party must conform to the IHR model SSC in Annex 3. The use of the model certificate facilitates the international movement of shipping, minimizes unnecessary delays, helps to standardize the inspection process, and allows uniform and easily recognizable communication of risks. Certificates must have correct formatting and content as specified in the IHR (2005);

1. According to the IHR (2005), "public health risk" is defined as "a likelihood of an event that may affect adversely the health of human populations, with an emphasis on one which may spread internationally or may present a serious and direct danger". This central concept underpins these guidelines and, together with other definitions, is important in understanding how the inspection process described here achieves the IHR (2005) goals.

certificates that do not conform to the model may be viewed by other competent authorities as invalid or may be invalid according to the IHR (2005).

Use of the template does not impose any obligation on the ship, other than those prescribed by the IHR (2005).

SSCs may be used as international communication tools and will usually be delivered in countries (or, sometimes, regions) other than the place of issue. Therefore, States Parties should usually issue and complete the certificates in English or in French.

2. Roles and responsibilities

States Parties shall take all practical measures consistent with the IHR (2005) to ensure that conveyance operators permanently keep conveyances for which they are responsible free from infection or contamination, including vectors and vector reservoirs (IHR [2005] Article 24). The application of control measures may be required if evidence of sources of infection or contamination is found. These measures may be carried out by the conveyance operator (either by the crew or by a private company under contract) or by the competent authority. The control measures applied should always be agreed upon and supervised by the competent authority (usually the port health authority).

The SSCs are designed to identify, assess and record any public health risks, and the consequent control measures that should be taken, while ships are in port. Public health risks are identified by epidemiological evidence, direct observation or measurement (or any combination of these). The competent authority should evaluate the risk in terms of the epidemiological situation and the severity of the risk. Control measures shall be applied at the point of entry, according to the conditions specified by the IHR (2005).

If clinical signs or symptoms of illness or disease and factual evidence of a public health risk (including sources of infection and contamination) are found on board a ship on an international voyage, the competent authority shall consider the ship as affected and may:

- (a) disinfect, decontaminate, disinsect or derat the conveyance, as appropriate, or cause these measures to be carried out under its supervision; and
- (b) decide in each case the technique employed to secure an adequate level of control of the public health risk as provided in these regulations. Where there are methods or materials advised by WHO for these procedures, these should be employed, unless the competent authority determines that other methods are as safe and reliable (IHR [2005] Article 27.1).

If the competent authority for the points of entry is not able to carry out the control measures required under IHR Article 27, the ship may nevertheless be allowed to depart, subject to the following conditions:

- (a) the competent authority shall, at the time of departure, inform the competent authority of the next known port of entry of the type of information referred to under subparagraph (b); and

• (b) in the case of a ship, the evidence found and the control measures shall be noted in the SSCC (IHR [2005] Article 27.2).

Therefore, ports should have the capacity to support the control measures adopted to prevent the spread of disease and disease agents. Such measures include cleaning, disinfection, decontamination, deratting and disinsection.

A port on the WHO list of ports authorized to issue SSCs should also have available trained personnel who can board ships, identify any significant risks to public health and take appropriate control measures. Thus, States should have nationally recognized training and competency requirements for public health or environmental health officers assigned to issue SSCs.

Subject to IHR (2005) Article 43, or as provided in applicable international agreements, ships shall not be refused *free pratique* by States Parties for public health reasons. In particular, they shall not be prevented from embarking or disembarking; discharging or loading cargo or stores; or taking on fuel, water, food and supplies. States Parties may subject the granting of *free pratique* to inspection and, if a source of infection or contamination is found on board, the carrying out of necessary disinfection, decontamination, disinsection or deratting, or other measures necessary to prevent the spread of the infection or contamination (IHR [2005] Article 28).

If the public health risk appears significant or if evidence of potential international spread of disease exists, the National IHR Focal Point and community-level health authorities should be alerted immediately by the competent authority.

2.1 Role of Competent Authority

The competent authority responsible for the implementation and application of health measures at points of entry is required, according to the IHR (2005) (Article 22), to:

• be responsible for monitoring baggage, cargo, containers, conveyances, goods, postal parcels and human remains departing and arriving from affected areas, so that they are maintained in such a condition that they are free from sources of infection or contamination, including vectors and reservoirs;
• ensure, as far as practicable, that facilities used by travellers at points of entry are maintained in a sanitary condition and are kept free from sources of infection or contamination, including vectors and reservoirs;
• be responsible for supervising any deratting, disinfection, disinsection or decontamination of baggage, cargo, containers, conveyances, goods, postal parcels and human remains, or sanitary measures for persons, as appropriate under these regulations;
• advise conveyance operators, as far in advance as possible, of their intent to apply control measures to a conveyance, and shall provide, where available, written information concerning the methods to be employed;
• be responsible for supervising the removal and safe disposal of any contaminated water or food, human or animal dejecta, wastewater and any other contaminated matter from a conveyance;

- take all practicable measures consistent with these regulations to monitor and control the discharge by ships of sewage, refuse, ballast water and other potentially disease-causing matter that might contaminate the waters of a port, river, canal, strait, lake or other international waterway;
- be responsible for supervising service providers for services concerning travellers, baggage, cargo, containers, conveyances, goods, postal parcels and human remains at points of entry, including the conduct of inspections and medical examinations as necessary;
- have effective contingency arrangements to deal with an unexpected public health event;
- communicate with the National IHR Focal Point on the relevant public health measures taken pursuant to these regulations.

2.2 Role of conveyance operators

According to the IHR (2005) (Article 24), States Parties must take all practicable measures consistent with the IHR to ensure that conveyance operators:

- comply with the health measures recommended by WHO and adopted by the State Party;
- inform travellers of the health measures recommended by WHO and adopted by the State Party for application on board;
- permanently keep conveyances for which they are responsible free from sources of infection or contamination, including vectors and reservoirs. The application of measures to control sources of infection and contamination may be required if evidence is found.

The ship's master must ensure that any cases of illness that are indicative of an infectious disease or evidence of a public health risk on board are relayed to the competent authority at port on arrival, as required by Article 28 and Annex 3.

The IHR (2005) (Annex 4) also requires that conveyance operators shall facilitate:

- inspections of the cargo, containers and conveyance;
- medical examinations of persons on board;
- application of other health measures under these regulations;
- provision of relevant public health information requested by the State Party.

Conveyance operators shall also provide to the competent authority a valid SSCEC or SSCC and a Maritime Declaration of Health (IHR [2005] Articles 37 and 39; Annexes 3, 4 and 8).

With regard to vector-borne diseases, the IHR (2005) Annex 5 provides specific measures applicable to conveyances and conveyance operators.

2.3 Role of inspecting officers

The role of the competent authority is to inspect areas, systems and services on board; verify the practical implementation of these systems and services; ascertain the sanitary condition of areas inspected; and recommend corrective actions or require measures to be taken when and where applicable. Any control measures required should be noted on the Evidence Report Form (see Annex 7) and will lead to the issue ance of an SSCC.

The IHR (2005) Annex 3 identifies areas, systems and services to be inspected on a conveyance. The IHR (2005) Annex 1 requires that States Parties should provide trained personnel at points of entry to inspect conveyances, to conduct inspection programmes and to ensure a safe environment for travellers using port facilities, as appropriate. In order to perform these duties, inspecting officers should demonstrate competency in the following areas before being assigned to inspection duties:

- Assessment of public health risks (including the effectiveness of systems implemented to control risks) by direct observation and by measurement with testing and sampling equipment. The assessment should be based on the information received from conveyance operators, agents or the ship's master, such as information from the Maritime Declaration of Health[2]; communications to port before a ship's arrival about public health events on board (IHR [2005] Article 28); traveller information; the disease status at port of origin, in transit and at the port of entry; and the application of personal protective techniques and related equipment.
- An understanding of the manner in which public health risks from microbiological, chemical and radiological agents affect human health and can be transmitted to individuals via other individuals, food, air, water, waste, vectors, fomites and the environment; appropriate measures to restrict radiation exposure to levels as low as reasonably practicable, if radiation risks are identified; and the protocol for seeking professional help for management of radiation risks and their effects.
- Use of operational procedures for notification, assessment and response, equipment and medicines; knowledge of environmental requirements relating to the size and type of conveyance; and knowledge of related, applicable guidelines (e.g. WHO, International Labour Organization [ILO], International Maritime Organization [IMO]).

3. Pre-inspection planning and administrative arrangements for issuing ship sanitation certificates

Before establishing an inspection programme, the inspection procedures and the administrative arrangements necessary for inspections and the issue of certificates should be in place.

2. The ship's master must ascertain the health status on board before arrival at port. The master must also deliver to the competent authority a complete Maritime Declaration of Health, which should be countersigned by the ship's surgeon (if one is on board), unless the competent authority does not require a declaration (include Article 37 and Annex 8; see WHO interim technical advice for case management of pandemics [H1N1] on ships).

3.1 General preparation and administrative arrangements for inspection of ships and issue of ship sanitation certificates

3.1.1 Communication

• Develop procedures for notification, assessment and response with regard to public health events on board (e.g. disease outbreaks, sources of infection and contamination, incidents and emergencies).
• Establish and maintain communication, reporting and tracking systems, in cooperation with other key agencies and departments such as the National IHR Focal Point and the national health surveillance system.
• Maintain an up-to-date, easily accessible list of ports authorized to issue SSCs; include an up-to-date list of contact details for the ports.
• Ensure that inspectors are capable of clear communication with ships' operators or agents and their crews.
• Determine and disseminate the correct information for communication to the competent authority, the conveyances and their operators or agents.
• Assess the volume, frequency and type of ship arrivals.

3.1.2 Training

• Develop and implement plans that identify training needs, qualification requirements and competency criteria.
• Ensure that inspectors can issue SSCs, including in English
• Familiarize inspectors with the checklists in this document.
• Familiarize inspectors with all necessary procedures for inspecting ships and issuing SSCs.
• Ensure that inspectors demonstrate appropriate knowledge of the types of certificates defined in this document and of the legal text of the IHR (2005).
• Train inspectors in the likely flow of inspections, according to the size and type of ships.

3.1.3 Equipment

• Ensure that the tools and equipment necessary for inspection and control measures, including personal protective equipment (PPE) and sampling equipment, are always available and in a good condition (see list of PPE and recommended equipment in Annexes 4 and 5).
• Ensure that the correct forms, the unique seals or stamps to authenticate certificates, and other administrative supplies are accessible and in good condition.

3.1.4 Administration

• Develop and implement a high-quality management scheme for monitoring, auditing and assessing the outcomes of inspections.
• Develop and implement a system for the administrative control and record management of issued SSCs; for example, establish and maintain a file system or secure database of inspections performed and certificates issued. The system should be able to identify previous deficiencies.

• Establish a system for collecting the agreed inspection fees.
• Identify the port areas required for the safe inspection of ships; the adoption of control measures, when applicable; and the facilities and services as listed in IHR (2005) Annex 1B.

3.2 Planning for on-site inspection

• Request and record the pre-arrival information supplied by the ship's operator or agent (i.e. confirm time of arrival, berth, request for inspection, previous and next port, health status on board, identity and contact details of conveyance operator or agent).
• Based on the received information, perform an assessment of public health risks; assemble the personnel and equipment necessary for the individual inspection.
• Prepare easy-to-understand information about the inspection procedures in a form that can be transmitted to the ship in advance (e.g. a leaflet). For unannounced inspections, this information should not be transmitted in advance. The information should contain:
- a list of the documents required for inspection;
- advice that a contact person must be available on board;
- advice that all areas must be accessible for inspection, in order that the inspection proceeds unhindered.

4. Measures and operational procedures for ship inspection and the issue of ship sanitation certificates

Inspections are designed to confirm that ships are operating in accordance with appropriate practices for assessment and control of health risks on board.

An inspection provides a snapshot of a ship's operations and the manner in which its systems are implemented and maintained. The inspecting officer should typically identify risks that arise from the activities on the ship, and the effectiveness of the ship's own assessment of risks and control measures. Both the quality of operating procedures and the extent of their implementation should be assessed. Specifically, the inspection should determine whether the ship's operator and/or master have identified relevant hazards, assessed health risks and identified suitable control measures to effectively manage those risks.

With regard to the issue of SSCs, a port should have appropriately trained personnel available to board a ship, identify any significant public health risks and order control measures. Before boarding, inspectors need to comply with all necessary administrative and technical procedures to ensure efficient and safe access to the ship. They should also follow procedures to ensure a reasonable level of safety while the ship is in dock, when boarding and during onboard inspections.

If a new certificate is to be issued, all areas must be inspected. The areas must be in a condition that prevents cross-contamination by the inspection process (e.g. as described in Annex 3).

Before an inspection begins, the ship's master should, if possible, be informed of the purpose of the inspection, be advised to prepare all the required documentation and be instructed to provide a contact person on board to facilitate the inspection.

An inspection usually includes a preliminary discussion with the ship's operator or agent and the master on matters relating to the ship's sanitation systems and procedures. Additionally, the relevant documentation sent to the master or representative by the competent authority before the inspection should be reviewed.

If a ship's risk assessment and risk management systems are unsatisfactory, evidence of implementation is inadequate, or unforeseen potential hazards are identified, the inspection officer should discuss these matters with the master at the conclusion of the visit. The discussion may include previous inspection reports, relevant current documentation and all food- and water-related activities undertaken on the ship.

Subsequently, a summary of the matters that do not comply with this document or with other associated technical documents (e.g. the WHO *Guide to ship sanitation*) should be confirmed in writing by the officer on the Evidence Report Form (see Annex 7). The officer should also note the relevant advice given, including an expected timeframe for any corrective action.

If control measures are noted on the existing SSCC or on an existing Evidence Report Form (see Annex 7), the inspection shall verify that these measures have been successfully implemented.

If the conditions under which the inspection and/or control measures are performed are incompatible with obtaining satisfactory results, the competent authority shall make a note to that effect on the SSCC.

4.1 Documentation review

For inspection purposes, information on the ship, its cargo and possible public health risks is needed, but the requested information and documents from conveyance operators need to be public health information necessary for these public health purposes.

With regard to health documents, the practice should follow the requirements in the IHR (2005) and documents listed in other international agreements, such as IMO conventions that relate to environmental protection and sanitation in general (e.g. the International Convention for the Prevention of Pollution from Ships 1973, as modified by the 1978 protocol [MARPOL]; and the Convention on Facilitation of International Maritime Traffic 1965, as amended 2006).

To help the ship's master prepare for the inspection, a list of all required documents should be sent in advance by the national authority (e.g. to the maritime agent) before the port health officer boards the ship for inspection.

The Maritime Declaration of Health (see model provided in IHR [2005] Annex 8) contains basic data relating to the state of health of crew and passengers during the voyage and on arrival at the port, and provides valuable information on:
- identification of the ship;
- ports of call within past 30 days (to be listed);
- all crew members and travellers within past 30 days (to be listed);
- validity of the existing SSC and whether re-inspection is required;
- affected areas visited.

The SSC (see model provided in IHR [2005] Annex 3) identifies all areas of public health risks and any required control measures to be applied.

The International Certificate of Vaccination or Prophylaxis (see model provided in IHR [2005] Annex 6) verifies that crew members and passengers have been vaccinated according to entry requirements.

The following documents, according to the document list in the Convention on Facilitation of International Maritime Traffic 1965 (as amended 2006), may be requested by the competent authority to assess public health risk:
- a General Declaration, to ascertain the ship's name, type and flag State; it also provides valuable information on the ship's requirements in terms of waste and residues, reception facilities and brief particulars of voyage;
- a Cargo Declaration and Ship's Stores Declaration, for information on the cargo (e.g. port of loading and discharge, description of goods);
- a Dangerous Goods Manifest, which details information on dangerous goods (e.g. subsidiary risk(s), mass, stowage position on board).

The following additional sources of information may be required, if appropriate, for assessment of public health risk:
- management plans concerning, for example, water bunkering, food safety, pest control, sewage or waste;
- an IMO Ballast Water Reporting Form;
- a medical log, for information on incidents on board the ship that may constitute health events under the IHR (2005);
- a list of medicines, providing information on the kinds and amounts of medicines carried in the medical chest;
- a Potable Water Analysis Report, which provides the results of any microbiological examinations or chemical analyses of potable water on board.

The competent authority should develop protocols and procedures for the transmission of pre-arrival and pre-departure information, to effectively process the required information.

4.2 Inspection process

Inspection is undertaken by observing areas of the ship. When a ship requests a new certificate, all areas as listed in this document (see Annex 3) should be inspected. The main purpose of the inspection is to confirm that all points of control have been correctly identified, and that any appropriate control measures have been implemented or corrective actions taken.

While performing inspections on board or in port areas, inspectors must wear appropriate identification, clothing and PPE, including, but not limited to, life jackets, safety helmets, safety boots, high-visibility clothing, respiratory and noise (ear) protection, rubber gloves, protection goggles, face masks (FFP3) and single-use overalls, as required.

Previous identification and security clearance of port authorities and ship operators should be granted before starting the inspection.

Generally, the inspector starts the inspection by introducing the team and outlining the objective of the inspection to the master. The inspector then receives information about operating conditions and safety rules on board from the master. This exchange should occur in a private space, if available. The inspection process is then outlined to the master, and the documentation in place is reviewed.

The order of inspection (see Annex 3) is at the inspector's discretion. However, cross-contamination from inspection activities should be avoided. Therefore, personal hygiene, cleanliness of clothes and the inspector's health status should be considered.

If a new certificate is to be issued, all areas have to be inspected. If the ship's holds are in use, the cargo should also be inspected, if applicable, especially for presence of vectors. If enough personnel are available in the inspection team, team members can be assigned different areas for inspection. The aim is to achieve all the objectives of the inspection, taking into consideration the availability of time, the number of inspectors, and the size and type of the ship.

The areas for inspection, the kind of evidence sought, the potential sources of information and the appropriate control measures to be taken are identified in the checklists of this handbook. This handbook can help identify deficiencies and non-compliance before completing the certificate.

4.3 Taking samples

The model SSC in IHR (2005) Annex 3 contains columns for recording "sample results" as part of the inspection and related information; however, such samples may not be required in all inspections according to the IHR. Whether a sample should be taken and analysed depends on factors such as the particular circumstances described in the checklists; the evidence found by the inspectors; the nature of any potential public health risks; and the adequacy, in a particular context, of the usual inspection techniques that do not involve taking samples. For example, if the cold potable water system shows temperatures above 25 °C, the risk of *Legionella* contamination increases. Therefore, this temperature is a trigger for taking a water sample.

Harmful contamination other than microbial contamination (e.g. from chemical or radioactive sources) may also be found on ships. Methods of sampling for these contaminants are described in the WHO *Guide to ship sanitation*.

If sample results are pending, issue an SSCC and note "Results are pending" on the certificate.

In general, when clinical signs or symptoms of illness or disease are present, evidence of a public health risk (including sources of infection and contamination) is found on board, or a public health risk is definitively identified, the competent authority determines the appropriate public health measures to be applied for an adequate level of control. Methods or materials advised by WHO for these measures should be employed, unless the competent authority determines that other methods are similarly safe and reliable.

4.4 Issue of certificates

SSCECs and SSCCs consist of two parts: (a) the model certificate, which outlines the key physical areas of the ship for inspection; and (b) the attached references to the systems for management of food, water, waste, swimming pools and spas, and medical and other facilities that may require closer inspection, according to the size and type of the ship. The Evidence Report Form can be used to list the evidence found and measures indicated.

After the inspection, the inspecting officer should debrief the master before issuing an SSC. The master or representative must be allowed sufficient time to address any deficiencies and to retrieve the necessary documentation before completing the certificate. According to the evidence of the adequacy of sanitary measures detected during the inspection, either an SSCEC or an SSCC is issued (see the flow chart in IHR [2005] Article 39, Appendix 2).

Instructions for completing the certificate are as follows:

- Strike through the non-applicable certificate in the heading (either SSCEC or SSCC).
- Fill in the required information in the two tables (name of ship, flag, etc.).
- Choose the applicable table (left: SSCEC, right: SSCC).
- Complete every box in all the columns.
- Write legibly and use consistent wording from the checklists of this handbook.
- Use the Evidence Report Form if there is insufficient space on the SSC.
- Note areas not applicable by marking "NA".
- Use the wording "None" or "Nil" in areas in which no evidence is found.
- List the documents reviewed.
- Use the wording "None" or "Nil" if no documents were reviewed.
- Clearly indicate if sample results were reviewed by noting "Yes" or "No".
- Indicate if sample results are not yet available by noting "Sample results pending".
- Sign (identifying the inspecting officer), date and stamp the certificate.
- Ensure that all certificates are legible.
- Ensure that English is at least one of the languages on the certificate.

If a re-inspection is performed on a ship holding a valid certificate, an Evidence Report Form should be attached to the original certificate to record further information. The attachment must be referenced on the original certificate, preferably with a stamp as shown in Section 4.4.4, with the signature of the inspector. The attachment should also refer to the original document.

4.4.1 Ship Sanitation Control Certificate

An SSCC is issued when evidence of a public health risk, including sources of infection and contamination, was detected on board and the required control measures have been satisfactorily completed. The SSCC records the evidence found, the control measures taken, and the samples taken and corresponding results (if applicable); if necessary, an Evidence Report Form can be attached.

If the conditions under which control measures are taken are such that, in the opinion of the competent authority, a satisfactory result cannot be achieved at the port where the operation was performed, the competent authority shall make a note to this effect on the certificate. The note identifies all evidence of ship-borne public health risks and any required control measures to be applied at the next port of call. If the ship is allowed to depart, at the time of departure the competent authority shall inform the next known port of call by a rapid means of communication (e.g. e-mail, fax, telephone) of the type of evidence and the requisite control measures. These procedures particularly apply in contexts in which the public health risk could spread internationally or could present a serious and direct danger to the health of human populations.

Any evidence of public health risks identified, required control measures or notation of sample results pending that could lead to the issue of an SSCC should be documented on the certificate, when applicable.

4.4.2 Ship Sanitation Control Exemption Certificate

According to the IHR (2005), an SSCEC is issued when no evidence of a public health risk is found on board and the competent authority is satisfied that the ship is free from infection and contamination, including vectors and reservoirs. This certificate is usually issued, as far as practicable, only if the inspection has been performed when the ship and the holds are empty, or when the holds contain only ballast or other material of similar nature and a thorough inspection of the holds is possible (IHR [2005] Article 39).

Despite the presence of a valid SSCEC or an extension, according to the IHR (2005), inspections may nonetheless be required in various circumstances, as stated in IHR (2005) Articles 23 and 27 and Annex 4 (e.g. if the pre-assessment indicates evidence of a public health risk).

4.4.3 Extension of ship sanitation certificates

SSCECs and SSCCs are valid for a maximum of six months. This period may be extended by one month if the inspection or control measures required cannot be accomplished at the port. However, if a ship constitutes a serious risk for the spread of disease, the necessary disinfection, decontamination, disinsection, deratting or other measures to prevent the spread of the infection or contamination must be performed at the next point of entry. At the time of departure, the competent authority shall inform the competent authority of the next point of entry of the evidence found and the control measures required.

An extension allows a ship to reach a port at which the inspection and necessary control measures can be performed, without the necessity to travel with an expired certificate.

An extension may be granted up to 30 days before the expiry date of the existing SSC. However, the SSC cannot be extended for longer than 30 days after the expiry date (IHR [2005] Article 39).

Use of an "extension stamp" similar to that shown below is recommended to ensure a common standard among competent authorities. Place the stamp on the existing certificate.

EXTENSION

The validity of this certificate has been extended till ___/___/_____ (dd/mm/yyyy) (max. 30 days after expiry date) by the competent authority in the **Port of XXXXXXXXX.**

Date and signature: ________________

4.4.4 Evidence Report Form

The Evidence Report Form (see Annex 7) can be used to document evidence of public health risks found during an inspection, and also the prescribed control measures or corrective actions. The words "required" and "recommended" are used, according to the evidence found, samples tested and documents reviewed. The inspecting officer then submits the SSC and attached Evidence Report Form to the ship's master. If a report form is used, a note is made on the SSC. Use of an "attachment stamp" similar to that shown below is recommended to ensure a common standard among competent authorities.

SEE ATTACHMENT

A document has been attached to this certificate by the competent authority in the **Port of XXXXXXXXX.**

This attachment consists of ____ pages

Date and signature: ___________________

Some control measures required to avoid dissemination of disease and to control an existing serious and direct danger should be adopted immediately. Any required control measure automatically results in the issue of an SSCC.

According to the risk assessment performed by the competent authority during the inspection, the crew and conveyance operator should follow any other recommendations for preventive measures to avoid potential risks.

Existing international standards and regulations are used as a baseline for defining the control measures advised as "required" or "recommended" measures. When "shall" is used to address a measure in the relevant articles and accounts of the international conventions, standards and regulations, the measure is deemed as a required measure. For example, the ILO (No. 68) Food and Catering (Ships' Crews) Convention specifies matters of food supply and catering arrangements designed to secure the health and well-being of ships' crews. This measure is addressed by "shall", and thus leads to a "required" designation.

Articles published in scientific journals provide evidence of public health risks that trigger public health emergencies at points of entry or in conveyances. The effects of such events justify specified measures to control contamination and infection. These articles provide a scientific basis to determine which measures should be required. For example, huge outbreaks of foodborne diseases that are often caused by pathogens resulting from improper temperature control in the food chain have been documented. Therefore, preventive measures concerning temperature control in food source, preparation, processing and service are crucial. Such measures are therefore identified as required measures.

Some measures draw on international best practices to achieve the goal of controlling infection and contamination in an effective and efficient manner.

4.4.5 Affected conveyances and ship sanitation certificates

According to IHR (2005) Articles 27 and 39, a conveyance shall be considered affected:

- if a valid SSCEC or SSCC cannot be produced;
- if clinical signs or symptoms of illness or disease are present and information based on fact or evidence of a public health risk exists, including sources of infection and contamination. In these circumstances, a conveyance is considered infected even if it possesses an SSCEC or an SSCC.

If the competent authority is unable to carry out the required control measures, or the results of control measures are ineffective, this should be clearly stated on an attachment to the certificate (e.g. an Evidence Report Form as in Annex 7). The attachment should detail the facts or evidence of public health risks and the required control measures. The attachment should be clearly marked as an attachment to the original certificate and cross-referenced by, for example, the date and port of issue (see Annex 2).

After a re-inspection of the ship in the next port to check that the control measures required by the previous competent authority have been performed, and that the measures have been verified as effective, notes to that effect must be made on the attachment. The ship shall then cease to be regarded as an affected conveyance under IHR (2005) Article 27, unless other public health risks were discovered on the re-inspection. The original validity date of the exemption certificate remains unaffected, unless a full inspection is performed and a new certificate issued, with the agreement of the master of the ship.

5. Control measures

Once the public health risks have been identified based on the information or evidence found, the competent authority should determine appropriate control measures and consider the adequacy of existing control measures. Public health risks can be controlled by a variety of means. The competent authority should enforce reasonable and practicable control measures, according to the risk assessment. Unnecessary or excessive measures should be avoided. In addition, the availability of technical resources and reasonable costs should be considered when assessing options for control.

When a public health risk exists, control measures that will reduce the risk to an acceptable level should be identified. The conveyance operator is responsible for controlling any onboard risks. Nevertheless, the competent authority should provide reasonable assistance to identify suitable and relevant control options.

Control measures for public health risks on ships should be applied only after all key parties (i.e. the master, the conveyance operator or agent and the port authorities involved in this activity) have been fully informed of the intended methods. Critical activities, such as the designation of port areas for quarantine of ships suspected of carrying a public health risk, should be identified well in advance, in cooperation with the port operator for ship movement. The schedule of work to be performed should be confirmed with the ship's supervisory crew members and noted in any corrective actions.

The methods suggested for detecting and measuring public health risks on ships (in Part B of this document) are based on information from existing guidelines, WHO State Party experts, international organizations and the shipping industry.

Disinsection, decontamination, deratting, disinfection and other sanitary procedures taken pursuant to the IHR (2005) shall be performed in a manner that avoids injury and, as far as possible, discomfort to persons. Also, environmental damage that affects public health, baggage, cargo, containers, conveyances, goods or postal parcels should be avoided (IHR [2005] Annex 4B.1). As far as practicable, facilities used by travellers at points of entry should be maintained in a sanitary condition and kept free from sources of infection or contamination, including vectors and reservoirs (IHR [2005] Article 22.2). These measures shall be initiated and completed without delay, and applied in a transparent and non-discriminatory manner (IHR [2005] Article 42).

6. Other relevant international agreements and instruments

While this document focuses on specific provisions of the IHR (2005), other international instruments and agreements also address related issues, such as crew safety and comfort, and some operational aspects such as facilitation, communication, maritime pollution, and safety and security of ships and ports. Such instruments and agreements include those adopted under the auspices of the ILO and the IMO. These international instruments should be compatible and, indeed, synergistic, with the IHR (2005). In any event, the IHR (2005) provides that the IHR and other instruments be interpreted in a compatible manner. A number of these instruments and agreements are referred to, where applicable, in parts of this document.

As provided in the IHR (2005) and subject to the preceding paragraph, the regulations do not prevent States Parties having certain common interests in terms of their health, geographical, social or economic conditions. States Parties are also not precluded from concluding special treaties or arrangements to facilitate the application of the regulations, with particular regard to:

- direct and rapid exchange of public health information between neighbouring territories of different States Parties;
- health measures to be applied to international coastal traffic and to international traffic in waters within their jurisdiction;
- health measures to be applied in contiguous territories of different States Parties at their common frontier;
- arrangements for carrying affected crew and passengers or affected human remains by means of transport specially adapted for the purpose;
- deratting, disinsection, disinfection, decontamination or other treatment designed to render goods free from disease-causing agents.

The IHR (2005) also provides that, without prejudice to their obligations under the regulations, States Parties that are members of a regional economic integration organization shall apply in their mutual relations the common rules in force in that organization.

PART B: CHECKLISTS FOR SHIP INSPECTION

Area 1 Quarters

Introduction

The operator is accountable for maintaining a safe environment on board for crew and passengers. As the IHR (2005) and other relevant international agreements should be interpreted in a compatible manner (Article 57), quarters for crew members should comply with existing conventions on crew accommodation and food and catering in the ILO conventions. For ships constructed before July 2006, all crew accommodation should comply with the Accommodation of Crews Convention (Revised) 1949 No. 92 and the Accommodation of Crews (Supplementary Provisions) Convention 1970 No. 133. For ships constructed after July 2006, accommodation should comply with the Maritime Labour Convention 2006.

International standards and recommendations

ILO Maritime Labour Convention 2006

1. Article IV, Seafarers' employment and social rights, paragraph 3: Every seafarer has a right to decent working and living conditions on board ship; paragraph 4: Every seafarer has a right to health protection, medical care, welfare measures and other forms of social protection;
2. Regulation 3.1, Accommodation and recreational facilities.

ILO Convention No. 92 concerning crew accommodation on board ship (revised 1949).
Accommodation of Crews Convention (Revised), 1949, lays down detailed specifications concerning matters as sleeping accommodation, mess and recreation rooms, ventilation, heating, lighting and sanitary facilities on board ship.

ILO Convention No. 133 on Accommodation of Crews 1970

ILO R140 Crew Accommodation (air Conditioning) Recommendation 1970

ILO Convention No. 147 on Minimum Standards in Merchant Ships 1976

Main risks

Factors contributing to the occurrence of public health risks on board include the design, construction, management and operation of quarters.

Document review

- Construction drawings of sanitary facilities and ventilation.
- Cleaning procedures and logs.
- Construction plans demonstrating how cross-contamination is avoided in specified clean and dirty areas.
- Smoke tests at exhaust and at air intakes close to exhaust.

References

International conventions

ILO, Maritime Labour Convention 2006.

Scientific literature

Barker J, Stevens D, Bloomfield SF (2001). Spread and prevention of some common viral infections in community facilities and domestic homes. *Journal of Applied Microbiology*, 91(1):7–21.

Black RE et al. (1981). Handwashing to prevent diarrhea in day-care centers. *American Journal of Epidemiology*, 113(4):445–451.
Carling PC, Bruno-Murtha LA, Griffiths JK (2009). Cruise ship environmental hygiene and the risk of norovirus infection outbreaks: an objective assessment of 56 vessels over 3 years. *Clinical and Infectious Diseases*, 49:1312–1317.
Centers for Disease Control and Prevention (2001). Influenza B virus outbreak on a cruise ship—Northern Europe, 2000. *Morbidity and Mortality Weekly Report*, 50:137–140.
Centers for Disease Control and Prevention (2002). Outbreaks of gastroenteritis associated with noroviruses on cruise ships—United States. *Morbidity and Mortality Weekly Report*, 51:1112–1115.
Centers for Disease Control and Prevention (2003). Norovirus activity—United States, 2002. *Morbidity and Mortality Weekly Report*, 52:41–45.
Chimonas MA et al. (2008). Passenger behaviors associated with norovirus infection on board a cruise ship—Alaska, May to June 2004. *Journal of Travel Medicine*, 15:177–183.
Corwin AL et al. (1999). Shipboard impact of a probable Norwalk virus outbreak from coastal Japan. *The American Journal of Tropical Medicine and Hygiene*, 61:898–903.
Cramer EH, Blanton CJ, Otto C (2008). Shipshape: sanitation inspections on cruise ships, 1990–2005, Vessel Sanitation Program, Centers for Disease Control and Prevention. *Journal of Environmental Health*, 70:15–21.
Depoortere E, Takkinen J (2006). Coordinated European actions to prevent and control norovirus outbreaks on cruise ships. *Euro Surveillance: European Communicable Disease Bulletin*, 11:E061018.
Enserink M (2006). Infectious diseases. Gastrointestinal virus strikes European cruise ships. *Science*, 313:747.
Hansen HL, Nielsen D, Frydenberg M (2002).Occupational accidents aboard merchant ships. *Occupational and Environmental Medicine*, 59(2):85–91.
Herwaldt BL et al. (1994). Characterization of a variant strain of Norwalk virus from a foodborne outbreak of gastroenteritis on a cruise ship in Hawaii. *Journal of Clinical Microbiology*, 32:861–866.
Ho MS et al. (1989). Viral gastroenteritis aboard a cruise ship. *The Lancet*, 2:961–965.
Lang L (2003). Acute gastroenteritis outbreaks on cruise ships linked to Norwalk-like viruses. *Gastroenterology*, 124:284–285.
Lawrence DN (2004). Outbreaks of gastrointestinal diseases on cruise ships: lessons from three decades of progress. *Current Infectious Disease Reports*, 6:115–123.
O'Neill HJ et al. (2001). Gastroenteritis outbreaks associated with Norwalk-like viruses and their investigation by nested RT-PCR. BMC *Microbiology*, 1:14.
Verhoef L et al. (2008). Multiple exposures during a norovirus outbreak on a river-cruise sailing through Europe, 2006. *Euro Surveillance: European Communicable Disease Bulletin*, 13(24)pii:18899.
Widdowson MA et al. (2004). Outbreaks of acute gastroenteritis on cruise ships and on land: identification of a predominant circulating strain of norovirus—United States, 2002. *Journal of Infectious Diseases*, 190:27–36.

Code of Areas	Inspection results: evidence found, sample results, documents reviewed	Control measures and corrective actions	Required	Recommended
1.1 Construction				
1.1.1 ❑	Surfaces and fixtures not properly sealed or difficult to clean.	Accommodation should be constructed of material that is easily sealed and cleaned.		❑
1.1.2 ❑	Accommodation constructed of unsuitable material that creates conditions for potential infestation by vectors.	All accommodation to be kept free from vector entry.		❑
1.1.3 ❑	No window available or ventilation system inadequate, which affects the health of occupants.	Adequate ventilation system or properly screened window should be in place, especially in sleeping rooms and mess rooms, to prevent the spread of disease. Adjust ventilation dependent on climate in which ship is sailing.		❑
1.1.4 ❑	No heating system.	Provide adequate heating system.		❑
1.1.5 ❑	Sleeping rooms for crew or passengers constructed from inferior materials.	Provide separate sleeping rooms constructed of steel or other approved substance; rooms must be watertight and gas-tight.		❑
1.1.6 ❑	No toilet facility available for crew members.	Provide toilet facility for crew members, either in their quarters or in a common lavatory outside individual quarters.	❑	
1.1.7 ❑	Bathrooms or shower rooms not provided.	Equip sleeping rooms with private or common bathroom, including toilet.		❑
1.1.8 ❑	Absence of drainage system or drainage system unable to cope with demand.	Install drainage system sufficient to cope with demand.		❑
1.1.9 ❑	External doors or windows not vector-protected.	Where appropriate, apply measures to ensure effective screening against vector entry.		❑
		Construct doors that open outward and are self-closing.		❑
		Provide informative material on individual preventive measures.		❑
1.1.10 ❑	Screen mesh not sufficiently small gauge (i.e. max.1.6 mm)	Install screen mesh of gauge 1.6 mm or less.		❑
1.1.11 ❑	Lack of ventilated space between toilet facilities, quarters and food spaces.	Modify construction of toilet facilities to enable open-air ventilation (prevents cross-contamination).		❑
1.1.12 ❑	Exhaust air vents from sanitary spaces physically connected to air supply systems, or the two systems are too close.	Correct the construction or design so that no exhaust air from sanitary or other spaces is connected to air supply systems.		❑

Code of Areas	Inspection results: evidence found, sample results, documents reviewed	Control measures and corrective actions	Required	Recommended
		1.2 Equipment		
1.2.1 ❑	Hand-washing facilities absent or inadequate.	Install proper hand-washing facilities (including liquid soap, paper towels, etc.). Provide informative material on personal hygiene, signage and a regular awareness programme for crew on use of toilet facilities and the importance of hand washing.		❑
1.2.2 ❑	Toilet facilities insufficiently equipped.	Equip toilet facilities with means for hand drying (preferably disposable paper towels), toilet paper and individual or liquid soap.	❑	
1.2.3 ❑	Inadequate storage of personal effects creating conditions for cross-contamination.	Provide adequate storage space for personal effects, or remove to individual quarters.		❑
		1.3 Cleaning and maintenance		
1.3.1 ❑	Cleaning and maintenance programme absent or inadequate.	Instigate adequate cleaning and maintenance programme.		❑
1.3.2 ❑	Poor hygiene conditions with presence of dust, waste, vectors.	Instigate cleaning and disinfection schedule.	❑	
1.3.3 ❑	Contamination by chemicals or other agents.	Apply decontamination measures.	❑	
1.3.4 ❑	Toilet leaks or overflows, or cross-connection exists.	Maintain toilet system free from leaks and backups.	❑	
1.3.5 ❑	Toilet flushing system poorly maintained.	Keep toilet flushing system maintained.		❑
1.3.6 ❑	Dirty linens and cloths.	Provide laundry facilities with appropriate equipment for laundry treatment, storage and distribution (clean and dirty circuits well defined).		❑
1.3.7 ❑	Evidence of vectors or reservoirs found.	Perform disinfection and appropriate disinsection or deratting measures.	❑	
		Repair or replace surfaces or fixtures to be durable, perform as originally designed, allow for easy cleaning and prevent vector infestation.		❑
		1.4 Lighting		
1.4.1 ❑	Natural or artificial lighting insufficient.	Provide artificial lighting when adequate natural light not available.		❑
		1.5 Ventilation		
1.5.1 ❑	Evidence of dirt and debris in heating or cooling systems, or poor air quality.	Relocate air-conditioning and heating systems to facilitate easy cleaning and disinfection.		❑

Area 2 Galley, pantry and service areas

Introduction

Major risk factors that contribute to foodborne outbreaks on board ships are primarily associated with temperature control of perishable food, infected food handlers, cross-contamination, heat treatment of perishable food, contaminated raw ingredients and use of non-potable water in the galley. Some diseases can be transmitted from one country to another by infectious agents or contaminants due to poor control measures on board. Therefore, detecting contamination in the sources, preparation and processing of food, as well as in the service of food at restaurants and in mess rooms, is crucial for the prevention and control of foodborne disease.

International standards and recommendations

Codex Alimentarius Commission (CAC)

The Codex Alimentarius is a collection of internationally adopted food standards created in 1963 by the Food and Agriculture Organization of the United Nations (FAO) and WHO. The standards in the collection are presented in a uniform manner. The Codex also includes advice in the form of codes of practice, guidelines and other recommended measures to assist in achieving the purposes of the Codex Alimentarius (CAC 1995; 1997a, b; 1999; 2003). The CAC guidance provides important information on basic food safety, which will be referred to throughout this section.

ILO, Maritime Labour Convention 2006

Regulation 3.2, Food and catering,paragraph 2: Each Member shall ensure that ships that fly its flag meet the following minimum standards: (b) the organization and equipment of the catering department shall be such as to permit the provision to the seafarers of adequate, varied and nutritious meals prepared and served in hygienic conditions; and (c) catering staff shall be properly trained or instructed for their positions.

The regulation also contains further requirements and guidance related to proper food handling and hygiene.

ILO (No. 68), Food and Catering (Ships' Crews) Convention 1946

Article 5: Each Member shall maintain in force laws or regulations concerning food supply and catering arrangements designed to secure the health and well-being of the crew of the vessels mentioned in Article 1.

These laws or regulations shall require:
(a) the provision of food and water supplies which, having regard to the size of the crew and the duration and nature of the voyage, are suitable in respect of quantity, nutritive value, quality and variety;
(b) the arrangement and equipment of the catering department in every vessel in such a manner as to permit of the service of proper meals to the members of the crew.

Article 6: National laws or regulations shall provide for a system of inspection by the competent authority of:
(a) supplies of food and water;
(b) all spaces and equipment used for the storage and handling of food and water;

(c) galley and other equipment for the preparation and service of meals; and
(d) the qualification of such members of the catering department of the crew as is required by such laws or regulations to possess prescribed qualifications.

Article 7: National laws or regulations or, in the absence of such laws or regulations, collective agreements between employers and workers shall provide for inspection at sea at prescribed intervals by the master, or an officer specially deputed for the purpose by him, together with a responsible member of the catering department of:
(a) supplies of food and water;
(b) all spaces and equipment used for the storage and handling of food and water, and galley and other equipment for the preparation and service of meals.
The results of each such inspection shall be recorded.

Hazard Analysis Critical Control Point system (HACCP)
HACCP is noted as a system to identify and monitor the critical control points in the food manufacturing and distribution chain, including the source and stockpile. At these critical points, control is essential to prevent, eliminate or reduce a hazard and to take corrective action. Food safety plans or programmes (FSPs) are required to manage the process of providing safe food. Typically, the FSP is based on the HACCP system.

Main risks
Foodborne diseases have been associated with loading poor-quality food. Nevertheless, even if the loaded food is safe, this does not ensure that the food will remain safe during the storage, preparation, cooking and serving activities that follow on board. The main risks to food safety in the galley, pantry and service areas are related to the following:
• Biological hazards (bacteria, viruses, fungi and parasites)
Biological hazards occur when bacteria, viruses, moulds, yeasts or parasites contaminate food. These organisms are commonly associated with humans and with raw products entering food preparation sites. Therefore, raw ingredients in the galley are high-risk factors. Storage time and temperature of food, and awareness and implementation of hygienic practices by food handlers on board ship also play significant roles in food safety.
• Chemical hazards (e.g. cleaning agents)
Chemical contamination of food may inadvertently occur "naturally" before loading or during processing (e.g. by the misuse of cleaning chemicals or pesticides). Examples of naturally occurring chemicals are mycotoxins (e.g. aflatoxin), scombrotoxin (histamine), ciguatoxin and shellfish toxins.
• Equipment and utensils
The equipment and utensils contacting food are designed and constructed to ensure that, when necessary, they can be adequately cleaned, disinfected and maintained to avoid the contamination of food. Equipment and containers are typically made of materials with no toxic effects when used as intended. Where necessary, equipment should be durable and movable, or capable of being disassembled to allow for maintenance, cleaning, disinfection, monitoring and inspection for pests.

Document review

- Cleaning schedule and logs.
- Purchase records and shipboard documentation of food sources (wrapping or other identification on the packaging, or a written product identification sheet).
- Food storage in–out record.
- Drainage construction drawings.
- Previous inspection reports.
- Pest logbook with information on sightings.
- Temperature records for food storage, cooling logs and thermometer readings.

References

International conventions

ILO, Maritime Labour Convention 2006.

Scientific literature

Addiss DG et al. (1989). Outbreaks of diarrhoeal illness on passenger cruise ships, 1975–85. *Epidemiology and Infection*, 103:63–72.

Berkelman RL et al. (1983). Traveler's diarrhea at sea: two multi-pathogen outbreaks caused by food eaten on shore visits. *American Journal of Public Health*, 73:770–772.

Boxman IL et al. (2009). Environmental swabs as a tool in norovirus outbreak investigation, including outbreaks on cruise ships. *Journal of Food Protection*, 72:111–119.

Cliver D (2009). Control of viral contamination of food and environment. *Food and Environmental Virology*, 1:3–9.

Couturier E et al. (2009). Cluster of cases of hepatitis A with a travel history to Egypt, September–November 2008, France. *Euro Surveillance: European Communicable Disease Bulletin*, 14(3) pii:19094.

Cramer EH, Blanton CJ, Otto C (2008). Shipshape: sanitation inspections on cruise ships, 1990–2005, Vessel Sanitation Program, Centers for Disease Control and Prevention. *Journal of Environmental Health*, 70:15–21.

Cramer EH, Gu DX, Durbin RE (2003). Diarrheal disease on cruise ships, 1990–2000: the impact of environmental health programs. *American Journal of Preventive Medicine*, 24:227–233.

Cramer EH et al. (2006). Vessel sanitation program environmental health inspection team. Epidemiology of gastroenteritis on cruise ships, 2001–2004. *American Journal of Preventive Medicine*, 30(3):252–257.

Herwaldt BL et al. (1994). Characterization of a variant strain of Norwalk virus from a foodborne outbreak of gastroenteritis on a cruise ship in Hawaii. *Journal of Clinical Microbiology*, 32:861–866.

Hobbs BC, Colbourne MJ, Mayner PE (1975).Food hygiene and travel at sea. *Postgraduate Medical Journal*, 51:817–824.

Lawrence DN et al. (1979). Vibrio parahaemolyticus gastroenteritis outbreaks aboard two cruise ships. *American Journal of Epidemiology*, 109:71–80.

Lew JF et al. (1991). An outbreak of shigellosis aboard a cruise ship caused by a multiple-antibiotic-resistant strain of Shigella flexneri. *American Journal of Epidemiology*, 134:413–420.

Mouchtouri VA et al. (2008). Surveillance study of vector species on board passenger ships, risk factors related to infestations. *BMC Public Health*, 8:100.

Rooney RM et al. (2004). A review of outbreaks of foodborne disease associated with passenger ships: evidence for risk management. *Public Health Reports*, 119(4):427–434.
Said B et al. (2009). Hepatitis E outbreak on cruise ship. *Emerging Infectious Diseases*, 15:1738–1744.
Sasaki Y et al. (2006). Multiple viral infections and genomic divergence among noroviruses during an outbreak of acute gastroenteritis. *Journal of Clinical Microbiology*, 44:790–797.
Snyder JD et al. (1984). Outbreak of invasive Escherichia coli gastroenteritis on a cruise ship. *American Journal of Tropical Medicine and Hygiene*, 33:281–284.
Waterman SH et al. (1987). Staphylococcal food poisoning on a cruise ship. *Epidemiology and Infection*, 99(2):349–353.

Guidelines and standards

WHO, Hazard Analysis Critical Control Point System (HACCP) (http://www.who.int/foodsafety/fs_management/haccp/en/)
Recommended international code of practice—general principles of food hygiene, CAC/RCP 1-1969, Rev. 4-2003. CAC

Code of Areas	Inspection results: evidence found, sample results, documents reviewed	Control measures and corrective actions	Required	Recommended
2.1 Documents and management practices review				
2.1.1 ❑	No food safety plan, written policies or informative material (documents and signage) on food preparation, handling and services in place.	Develop and implement food safety plan, based on HACCP principles, in terms of source, preparation, service, roles and responsibilities.		❑
		Post written policies and informative material (documents and signage) on food handling and production, hand washing and hygiene in a noticeable place in or near the galley.		❑
2.1.2 ❑	Medical logs indicate that crew affected by communicable diseases returned to work in the galley before being symptom free for a minimum of 48 hours, or the presence of other communicable diseases affecting the crew.	Food handlers or galley crew members with symptoms of gastrointestinal illness must not perform any food-related work until symptom free for a minimum of 48 hours.	❑	
		Re-evaluate communicable disease status.	❑	
2.1.3 ❑	No routine cleaning programme and schedule.	Develop written policies for hygiene, cleaning and maintenance procedures. Provide informative material (e.g. documents, videos, text books, signage) for crew.	❑	
		Develop training manual and improved supervision of programme application.	❑	
2.1.4 ❑	No temperature logs for received goods, freezers, cold storage, holding temperatures or preparation temperatures. No calibrated thermometers available.	Set up temperature logs for freezers and hot and cold holding units. Keep temperature control and cooking time calibration logs of food thermometers.	❑	
2.1.5 ❑	No waste management plan or cleaning schedule available.	Develop and implement waste management plan to prevent odour and nuisances, minimize attraction of vectors, avoid contamination of food and pollution of the environment. Set up cleaning schedule and logs.	❑	
2.1.6 ❑	No food safety training programmes or documented evidence that crew have undergone training.	Formulate and implement training programme and set up training log.		❑
2.2 Equipment, utensils and materials				
2.2.1 ❑	Hand-washing station in the galley absent or inadequately equipped.	Equip at least one dedicated hand-washing station, preferably in the galley area, with soap, means for hand drying (preferably disposable paper towels) and waste towel receptacle.	❑	
		Post signage that indicates location of hand-washing station, and proper hand-washing technique and time.	❑	

Code of Areas	Inspection results: evidence found, sample results, documents reviewed	Control measures and corrective actions	Required	Recommended
2.2.2 ❑	Muiti-use sink used for food preparation without proper cleaning and disinfection.	Set up food preparation sinks in as many areas as necessary and feasible—that is, in all meat, fish and vegetable preparation rooms; cold pantries or grade mangers; and any other areas where crew wash or soak food. Provide at least one sink dedicated to food preparation only.	❑	
		Strict cleaning, disinfection and sanitizing of sink before food preparation, especially if only one sink available.	❑	
2.2.3 ❑	Food contact surfaces, utensils and equipment not durable, corrosion resistant and non-absorbent.	Replace materials of food contact surfaces with corrosion-resistant, non-toxic, non-absorbent, easily cleanable, smooth, durable materials.	❑	
2.2.4 ❑	Tight-fitting doors or similar protective closures not available or not functioning.	Install or repair tight-fitting doors or similar protective closures for openings to ice bins, food display cases, and other food- and ice-holding facilities to prevent contamination of stored products.		❑
2.2.5 ❑	Inadequate waste containers (e.g. not rodent proof, watertight or non-absorbent; difficult to clean).	Use material and containers that are rodent proof, watertight, non-absorbent and easy to clean.		❑
2.2.6 ❑	Lids and covers on waste containers absent or not kept closed.	Keep lids and covers in food-handling spaces closed as much as possible during food preparation, food serving and cleaning operations.	❑	
		2.3 Facilities		
2.3.1 ❑	Evidence found of non-potable water use in galley, pantry and food stores.	Connect sinks with the potable water system.	❑	
2.3.2 ❑	Construction of areas, surfaces and equipment makes cleaning difficult, and allows vectors to harbour and food residues to build up.	Make areas, surfaces and equipment of materials that are durable and easy to clean and allow proper drainage.	❑	
2.3.3 ❑	Absent or inappropriate facilities for storing potable water and ice for use in food and drinks.	Ensure appropriate facilities for storage and distribution of an adequate supply of potable water.	❑	
2.3.4 ❑	Sink drain directly connected to wastewater system.	Reconstruct sink drain with an indirect connection to wastewater system (i.e. an air break) to prevent wastewater backup.	❑	
2.3.5 ❑	Absent or inadequate ventilation system (e.g. excessive condensation).	Redesign, reconstruct, adequately maintain and clean ventilation systems. Ensure louvres or registers in ventilation terminals are readily removable for cleaning.	❑	
2.3.6 ❑	Insufficient lighting levels.	Provide artificial lighting if adequate natural light not available to properly evaluate sanitary conditions.	❑	

Code of Areas	Inspection results: evidence found, sample results, documents reviewed	Control measures and corrective actions	Required	Recommended
2.3.7 ❑	Unprotected lighting over food preparation areas.	Install or fix lighting fixtures to ensure food is not contaminated by glass breakage.	❑	
2.3.8 ❑	Toilet facilities for food-handling crew absent or not accessible.	Provide dedicated, appropriate and hygienic spaces with toilet, hand-washing and hand-drying facilities, and an adequate supply of soap. Ensure facilities do not open directly into galleys or other food-handling areas.		❑
		Equip adequate changing facilities for food-handling crew; include suitable storage facilities for clothes, if possible.		❑
2.3.9 ❑	Dining room station(s), decks and areas under equipment or in technical places are not constructed of hard, durable, non-absorbent materials.	Equip dining room service stations with hard, durable, non-absorbent deck (e.g. sealed granite or marble). Ensure a safe separation distance of at least 61 cm (2 ft) from the edge of the working sides of the service station. Decks behind service counters, under equipment and in technical spaces must be constructed of hard, durable, non-absorbent materials (e.g. tiles, epoxy resin or stainless steel).	❑	
		Durable coving with radius at least 10 mm (0.4 in), or open design >90°, must be used as an integral part of the deck and bulkhead interface, and at the juncture between decks and equipment foundations. Stainless steel or other coving, if installed, should be securely installed and of sufficient thickness to be durable.	❑	
2.3.10 ❑	Food contact surfaces not smooth or have breaks, open seams, cracks, chips, inclusions, pits or other imperfections; or have sharp internal angles, corners and crevices; or are not easily accessible for cleaning and inspection.	Repair or replace damaged surfaces and equipment to ensure they are non-toxic, durable, corrosion resistant, non-absorbent, smoothly finished and easy to clean, to prevent cross-contamination.	❑	
		Clean, apply sanitizing measures, and disinsect or derat where vectors present.	❑	
2.4 Food processing				
2.4.1 ❑	Evidence of lack of knowledge of hand hygiene found (e.g. same employee loads dirty dishes and removes clean dishes, but does not thoroughly wash hands between the two tasks).	Wash hands regularly, especially between different tasks.	❑	
2.4.2 ❑	Evidence of cross-contamination between cooked and raw food.	Use separate utensils and cutting boards for preparation of raw and ready-to-eat foods. Clean and disinfect food contact surfaces, tableware and utensils whenever there is a change of use from raw to ready-to-eat foods.	❑	
		Clean and disinfect the cross-contamination area before food preparation. Separate preparation and storage areas for raw and ready-to-eat foods.	❑	

Code of Areas	Inspection results: evidence found, sample results, documents reviewed	Control measures and corrective actions	Required	Recommended
2.4.3 ❑	Foods in freezers visibly spoiled, refrozen or, when hand pressure is applied, partially thawed.	Repair or replace freezers that cannot maintain foods in a frozen state.	❑	
		Discard any spoiled, refrozen, partially thawed and thawed foods, and any other food stored at incorrect temperatures.	❑	
2.4.4 ❑	Perishable foods found stored at incorrect temperatures for the type or class of food. If time control used, no explanation or documentation for periods longer than 6 hours.	Maintain correct temperatures for storing perishable foods to prevent food entering the danger zone for microbial growth, as follows: • Place food held hot in a hot-holding apparatus already at a temperature of at least 62.8 °C (145 °F) and maintain at that temperature until required. • Reheat cooked, cooled, perishable food for hot-holding until all parts of the food reach a temperature of at least 74 °C (165 °F). Maintain temperature until required. • Store perishable foods and drinks at, or below, 4 °C (40 °F), except during preparation or when held for immediate serving after preparation. When such foods are stored for extended periods, a temperature of 4 °C (40 °F) is recommended. • Usually store fruits and vegetables in cool rooms.	❑	
2.4.5 ❑	First-aid box not accessible.	Ensure first-aid box is readily accessible for use in food-handling areas, and trained crew are appointed to take charge of first-aid incidents.		❑
2.4.6 ❑	Food handlers or galley crew members have exposed cuts and wounds.	Treat wounds with waterproof dressings. Wear disposable gloves if wounds become infected; apply medical treatment immediately.	❑	
2.5 Hygiene control system				
2.5.1 ❑	Temperature logs for hot- and cold-holding units not up to date, or inspection verifies temperature logs inaccurate.	Monitor temperature in hot- and cold-holding units on a regular basis. Maintain temperature logs and record any deviations accurately.		❑
2.5.2 ❑	Ambient air temperature and food probe thermometers absent or malfunctioning.	Provide at least one air temperature thermometer in cold-holding unit, and ensure at least one accurate food probe thermometer is used in the galley.		❑
2.5.3 ❑	Expired food.	Discard expired food and check all expiry dates on a regular basis.	❑	

Code of Areas	Inspection results: evidence found, sample results, documents reviewed	Control measures and corrective actions	Required	Recommended
		2.6 Personal hygiene		
2.6.1 ❑	Evidence of poor personal hygiene practices.	Strictly clean hands when personal cleanliness may affect food safety: •at the start of food-handling activities; •immediately after using the toilet; •between handling raw food or any material that could contaminate other food items and handling ready-to-eat food (avoid if possible); •between handling money and handling ready-to-eat food (avoid if possible). Refrain from: •smoking in or near food preparation and serving areas; •spitting in or near food preparation and serving areas; •chewing or eating in or near food preparation and serving areas; •sneezing or coughing over unprotected food. Always wear clean clothes.	❑	
2.6.2 ❑	Food handlers or galley crew members with signs or symptoms of communicable diseases (discharge from nose, eyes or ears; cough; diarrhoea; vomiting; fever; visibly infected skin lesions or boils; jaundice, etc.).	Any galley crew members showing signs or symptoms of communicable diseases must immediately report to the assigned medical officer. Food handlers or galley crew members should refrain from any food-related work until symptom free for a minimum 48 hours.	❑	
		2.7 Cleaning and maintenance		
2.7.1 ❑	Construction of, and materials used for, equipment and other galley features conducive to the build-up of food debris, grease and soil; galley fixtures not made of non-corroding metal or other durable material; or fixtures not close fitting.	Repair or replace galley surfaces, equipment and fixtures to be durable, close fitting and easily cleaned, and to prevent contamination of food and harbouring of vectors. Clean or replace equipment showing signs of grease or soil accumulation.		❑
		Implement a routine maintenance schedule for fixtures, fittings and equipment used during food production and food handling.		❑
2.7.2 ❑	Evidence of improper cleaning procedures and improper use of cleaning chemicals and disinfectants.	Handle, use and store cleaning chemicals carefully, in accordance with manufacturers' instructions.		❑
		Implement a checklist system for scheduled cleaning of all necessary items.		❑
2.7.3 ❑	Waste containers are a source of contamination and attract vectors.	After each emptying, thoroughly scrub, wash and treat waste containers with disinfectant.		❑
2.7.4 ❑	Evidence of accumulated soil and grease on previously cleaned food contact surfaces.	Properly clean and disinfect food contact surfaces, tableware and utensils after each use.	❑	

Code of Areas	Inspection results: evidence found, sample results, documents reviewed	Control measures and corrective actions	Required	Recommended
2.7.5 ❑	Evidence of inadequate cleaning and sanitizing of utensils or areas before using.	Use warm, soapy water and designated clean cloths to clean and thoroughly rinse utensils or areas after use. Additionally, apply approved chemical sanitizer at correct concentration, where appropriate.	❑	
2.7.6 ❑	Openings evident between decks and bulkheads; decks damaged or corroded.	Tighten bulkhead joints to prevent contamination of food and vector infestation.	❑	
		Repair or replace decks to be hard, durable, non-absorbent and non-slip.		❑
		Repair or replace any junctions between fixtures, decks and ceilings where openings allow for vector entry (use temporary measures, if needed, to close openings until appropriate permanent construction can take place).	❑	
2.7.7 ❑	Evidence of vector infestation.	Clean and disinfect food contact surfaces and apply vector control measures.	❑	
		Keep record of traps, baits (location, dates and results) and potential breeding sites for mosquitoes and other pests.	❑	
2.7.8 ❑	Evidence of vectors feeding or breeding inside or outside waste containers.	Apply disinfection and vector control measures.	❑	
		2.8 Food hygiene training		
2.8.1 ❑	Food handlers do not demonstrate competencies concerning hygiene.	Strengthen training of food handlers; first-level training should cover: •types and sources of public health risks related to the food chain; •basic knowledge of microbiology, toxins, spores, including growth and destruction of food contaminants; •food operation areas and equipment; •personal hygiene (basic rules and responsibilities); •preventing food contamination and spoilage; •cleaning, disinfection and sterilization; •legal obligations; • pest awareness; •effective temperature control of food, including: -chilled or frozen food -storage, thawing and cooking food -cooling, reheating and holding food; •common food hazards, including physical, chemical and microbiological hazards; symptoms and causes of food poisoning.		❑
2.8.2 ❑	Managers and supervisors of food processing demonstrate lack of necessary knowledge of food hygiene principles and practices.	Improve training of managers and supervisors to enable them to judge potential risks and take necessary action to remedy deficiencies.		❑

Appendix 1 Examples of appropriate temperatures and conditions for foods supplied to ships

Item	Temperature on receiving	Condition on receiving
Meat and poultry	5 °C (41 °F) or below.	Obtained from an approved source (i.e. stamped with official inspection stamp). Good colour and no odour. Packaging clean and in good condition.
Seafood	5 °C (41 °F) or below. The Codex recommends a temperature as close as possible to 0 °C.	Obtained from an approved source. Good colour and fresh odour. Packaging clean and in good condition.
Shellfish	7 °C (45 °F) or below. The Codex recommends a temperature as close as possible to 0 °C.	Obtained from an approved source. Clean, shells closed, no broken shells. Shellstock tags must be readable and attached.
Crustacea (unprocessed)	7 °C (45 °F) or below.	Obtained from an approved source. Clean and in good condition.
Crustacea (cut or processed)	5 °C (41 °F) or below.	Obtained from an approved source. Clean and in good condition.
Dairy products	5 °C (41 °F) or below, unless labelled otherwise.	Obtained from an approved source. Packaging clean and in good condition.
Shell eggs	7 °C (45 °F) or below.	Obtained from an approved source. Clean, not cracked.
Liquid eggs	5 °C (41 °F) or below.	Obtained from an approved source. Liquid eggs frozen and pasteurized.

Appendix 2 General principles of temperature control

Process	Operation		Temperature control	Required	Recommended
Thawing	a)	In refrigerator or purpose-built thawing cabinet	4 °C or below.	❑	
	b)	In running potable water	Not above 21 °C for a period not exceeding 4 hours.	❑	
	c)	In commercial microwave oven	Only when the food will be immediately transferred to conventional cooking units as part of a continuous cooking process, or when the entire, uninterrupted cooking process takes place in the microwave oven.	❑	
Cooking	a)	Rare cooked beef	The centre of joints must reach a minimum of 63 °C.	❑	
	b)	Large poultry carcasses	Temperature of 74 °C is achieved in the deep thigh muscle.	❑	
	c)	Milk (Code of Hygienic Practice for Milk and Milk Products, CAC/RCP 57, 2004)	72°C for 15 seconds (continuous flow pasteurization), or 63 °C for 30 minutes (batch pasteurization).	❑	
Portioning	a)	Chilled product	Completed within the minimum practicable time, which should not exceed 30 minutes.	❑	
			In large-scale systems in which cooking and chilling of foods cannot be performed in 30 minutes, portioning should take place in a separate area in which the ambient temperature should be 15 °C. Serve food immediately or place in cold storage at 4 °C.	❑	
Chilling and storage of chilled food	a)	Chilling	Reduce temperature in the centre of the food product from 60 °C to 10 °C in less than 2 hours. Immediately store product at 4 °C.	❑	
	b)	Storage	As soon as the chilling is complete the products should be put into a refrigerator. Temperature should not exceed 4 °C in any part of product and should be maintained until final use. Storage period between preparation of chilled food and its consumption should be less than 5 days, including both day of cooking and day of consumption.	❑	
Freezing and storage conditions for frozen food	a)	Freezing	Kept at or below –18 °C.	❑	
	b)	Storage	Stored at or below 4 °C for less than 5 days. Do not refreeze thawed or partially thawed food.	❑	

Process		Operation	Temperature control	Required	Recommended
Transport	a)	Vehicles and containers intended for transporting heated food.	Designed to maintain food temperature of at least 60 °C.	❑	
	b)	Vehicles and containers intended for transporting cooked-and-chilled food.	Designed to maintain temperature of already chilled (cooked) food, not to chill food. Ideally, maintain temperature of foods at 4 °C, but temperature may rise to 7 °C for a short period during transport.	❑	
	c)	Vehicles and containers intended for transporting cooked-and-frozen food.	Maintain at or below –18 °C, but may rise to –12 °C for a short time during transport.	❑	
Reheating and service	a)	Reheating.	At least 75 °C should be reached in centre of food within 1 hour of removing food from refrigeration.	❑	
			Reheat rapidly to pass food quickly through the hazardous temperature range between 10 °C and 60 °C.	❑	
	b)	Service.	Serve reheated food to consumer as soon as possible and at a temperature of at least 60 °C.	❑	
			In self-service establishments, maintain temperature of food either below 4 °C or above 60 °C, as appropriate.	❑	

Area 3 Stores

Introduction

The above risk factors applicable to galleys also apply to the food stores, as food stores are not only areas for food storage, but also contain counters and equipment for food or drink preparation, limited cooking and ware washing used for temporary food storage.; often, however, counters and equipment for food or drink preparation, limited cooking and ware washing are used for temporary food storage. Review all galley sections above for application to pantries of standards, evidence and corrective actions in all general categories of food safety.

Non-food stores include spaces designed for storage of non-food items, such as cleaning equipment, chemicals and other non-food equipment or supplies for support of food areas. Appropriately and hygienically managed stores limit potential for soiling of clean supplies and attraction and breeding of pests (i.e. non-food stores should be clean, organized, well stocked and well run). Storage areas should be appropriately labelled.

International standards and recommendations

ILO (No. 155), Occupational Safety and Health Convention 1981

Article 7: The situation regarding occupational safety and health and the working environment shall be reviewed at appropriate intervals either over-all or in respect of particular areas, with a view to identifying major problems, evolving effective methods for dealing with them and priorities of action, and evaluating results.

ILO (No. 134), Prevention of Accidents (Seafarers) Convention 1970

Article 4: These provisions shall refer to any general provisions on the prevention of accidents and the protection of health in employment which may be applicable to the work of seafarers, and shall specify measures for the prevention of accidents which are peculiar to maritime employment.

ILO (No. 68), Food and Catering (Ships' Crews) Convention 1946

Article 5: Each Member shall maintain in force laws or regulations concerning food supply and catering arrangements designed to secure the health and well-being of the crews of the vessels mentioned in Article 1.
These laws or regulations shall require:
(a) the provision of food and water supplies which, having regard to the size of the crew and the duration and nature of the voyage, are suitable in respect of quantity, nutritive value, quality and variety;
(b) the arrangement and equipment of the catering department in every vessel in such a manner as to permit of the service of proper meals to the members of the crew.

Article 6: National laws or regulations shall provide for a system of inspection by the competent authority of:
(a) supplies of food and water;
(b) all spaces and equipment used for the storage and handling of food and water;
(c) galley and other equipment for the preparation and service of meals; and
(d) the qualification of such members of the catering department of the crew as is required by such laws or regulations to possess prescribed qualifications.

Article 7: National laws or regulations or, in the absence of such laws or regulations, collective agreements between employers and workers shall provide for inspection at sea at prescribed intervals by the master, or an officer specially deputed for the purpose by him, together with a responsible member of the catering department of:
(a) supplies of food and water;

(b) all spaces and equipment used for the storage and handling of food and water, and galley and other equipment for the preparation and service of meals.
The results of each such inspection shall be recorded.

Codex Alimentarius Commission (CAC)
The Codex Alimentarius is a collection of internationally adopted uniform food standards. It also includes provisions of an advisory nature in the form of codes of practice, guidelines and other recommended measures to assist in achieving the purposes of the Codex Alimentarius (CAC 1995; 1997a, b; 1999; 2003). The CAC guidance provides important information on basic food safety, which will be referred to throughout this section.

Hazard Analysis Critical Control Point system (HACCP)
HACCP is noted as a system to identify and monitor the critical control points in the food manufacturing and distribution chain, including the source and stockpile. At these critical points, control is essential to prevent, eliminate or reduce a hazard and to take corrective action. Food safety plans or programmes (FSPs) are required to manage the process of providing safe food. Typically, the FSP is based on the HACCP system.

Main risks

The major risks include bacteria, viruses, fungi and parasites in, or on, food due to the improper storage of foods. For example:

- foods placed on the deck;
- improper holding temperatures in cold stores;
- eggs, fish, meat and poultry not separated from ready-to-eat foods (e.g. lunch meats from cut melons, salads and other ready-to-eat foods);
- washed and unwashed fruits and vegetables not separated.

Improper storage of chemicals is another risk.

Document review

- Cleaning and maintenance schedule and logs.
- Purchase records and shipboard documentation of food source (e.g. wrapping or other identification on packaging, or written product identification sheets).
- Food storage in–out records.
- Construction drawings
- Previous inspection reports.
- Pest logbook with information on sightings.
- Records of food storage temperatures, cooling logs and thermometer readings.

References

International conventions
ILO, Maritime Labour Convention 2006.

Scientific literature
Cramer EH, Gu DX, Durbin RE (2003). Diarrheal disease on cruise ships, 1990–2000: the impact of environmental health programs. *American Journal of Preventive Medicine,* 24:227–233.
McEvoy M et al. (1996). An outbreak of viral gastroenteritis on a cruise ship. *Communicable Disease Report CDR Review,* 6:R188–R192.
Rooney RM et al. (2004). A review of outbreaks of foodborne disease associated with passenger ships: evidence for risk management. *Public Health Reports,* 119:427–434.

Guidelines and standards
Codex Alimentarius Commission (http://www.codexalimentarius.net/web/index_en.jsp)
WHO, HACCP (Hazard Analysis Critical Control Point System)
(http://www.who.int/foodsafety/fs_management/haccp/en/)

Code of Areas	Inspection results: evidence found, sample results, documents reviewed	Control measures and corrective actions	Required	Recommended
		3.1 Construction		
3.1.1 ❑	Poorly designed construction for protection against weather and sea.	Reconstruct to ensure protection against weather and sea, insulation from heat or cold and separation from other spaces.	❑	
		Reconstruct to ensure that room is visually clean and structurally sound.	❑	
3.1.2 ❑	Openings or damage.	Repair openings and areas of significant damage.	❑	
		3.2 Cleaning and maintenance		
3.2.1 ❑	Soiled stores.	Maintain cleaning programmes and logs.	❑	
3.2.2 ❑	Evidence of standing water.	Eliminate standing water and its source.	❑	
3.2.3 ❑	Evidence of vectors or reservoirs.	Perform disinfection, disinsection and deratting measures.	❑	
		3.3 Food sources		
3.3.1 ❑	Foods found spoiled or unpackaged. Containers or packaging have no source or suspicious source identifications. Supervisory food crew members cannot provide satisfactory details of sources and countries of origin to enable tracing if a poisoning occurs	Document on food packaging or separately record sources to comply with country-of-origin laws and regulations.	❑	
		Obtain all foods consumed on board from reputable sources ashore (i.e. sources approved or considered satisfactory by the relevant health administration).	❑	
		Verify food quality and safety when purchasing: clean, free from spoilage and adulteration, safe for human consumption. Do not accept raw materials and ingredients if known to contain parasites; undesirable microorganisms, pesticides, veterinary drugs or toxins; decomposed or extraneous substances; unless contaminants can be reduced to an acceptable level by routine sorting or processing.	❑	
		Discard spoiled foods.	❑	
		3.4 Storage		
3.4.1 ❑	Food and non-food, or raw and prepared products not separate.	Separate food and non-food stores. Clearly segregate storage of raw and prepared products.	❑	

Code of Areas	Inspection results: evidence found, sample results, documents reviewed	Control measures and corrective actions	Required	Recommended
3.4.2 ❑	Disordered stores.	Post signage for storage and maintenance procedures.		❑
3.4.3 ❑	Foods found in contact with the deck, standing water or other contamination.	Store food at safe distance (approximately 15 cm or 6 in) above the deck and protect from water entry and other potential contamination.	❑	
		Dispose of contaminated foods and, where feasible, clean and sanitize food containers.	❑	
3.4.4 ❑	Storage areas poorly constructed or maintained, allowing presence of vectors or soiling of stored foods with dirt, debris or droppings.	Stock raw materials and ingredients in an order for effective stock rotation.		❑
		Reconstruct food storage areas with suitable materials for easy cleaning to deter harbouring vectors.		❑
		Repair or construct decks with hard, durable, non-absorbent, non-skid material. Install durable coving with a radius of at least 10 mm (0.4 in), or an open design >90°. Make the coving an integral part of the deck and bulkhead interface, and of the juncture between decks and equipment foundations, for easy cleaning and prevention of vector entry.		❑
		Apply disinsection and deratting measures to eliminate evident vectors.	❑	
3.4.5 ❑	Foods stored in locker rooms, toilet or bathing areas, garbage rooms, or mechanical or technical spaces; or under sewer lines, leaking water lines or lines on which water has condensed.	Always store foods in designated, secured rooms, protected from contamination and infestation.		❑
		Transfer foods to rooms free from contamination or temperature abuse for safe storage.	❑	
3.4.6 ❑	Perishable foods found stored under inadequate temperature conditions for the type or class of food, for periods over 4 hours, without sufficient explanation or documentation (e.g. cooling without a cooling log). Frozen foods in freezers found visibly spoiled or partially thawed when hand pressure applied	Maintain correct temperatures for storing perishable foods to prevent food entering the danger zone for microbial growth, as follows: • Place food held hot in a hot-holding apparatus already at a temperature of at least 62.8 °C (145 °F), and maintain at that temperature until required. • Reheat cooked, cooled, perishable food for hot-holding until all parts of the food reach a temperature of at least 74 °C (165 °F). Maintain temperature until required. • Store perishable foods and drinks at or below 4 °C (40 °F), except during preparation or when held for immediate serving after preparation. When such foods are stored for extended periods, a temperature of 4 °C (40 °F) is recommended. • Usually store fruits and vegetables in cool rooms.		❑

Code of Areas	Inspection results: evidence found, sample results, documents reviewed	Control measures and corrective actions	Required	Recommended
		Discard perishable foods immediately. If control measures applied in presence of inspector, note both date that control measures applied and re-inspection date (i.e. the same date) on the ship sanitation certificate.	❑	
3.5 Hazardous material				
3.5.1 ❑	Chemicals used for cleaning and maintenance of food areas stored in food area.	Separate chemicals by storing in a locker.	❑	
3.6 Training				
3.6.1 ❑	Evidence of absence or inadequate knowledge of correct use of cleaning chemicals.	Improve training in correct use of cleaning chemicals.		❑

Area 4 Child-care facilities

Introduction

Infants and children are known reservoirs for infection. Therefore, child-care facilities on board ships contribute to public health risks. They are also important centres for surveillance and control of public health risks. Modes of infectious disease transmission within or from child-care facilities include the droplet–air and oral–faecal routes and person-to-person spread. The prevalence of disease may depend on the level of immunity in children and carers, country of origin and age of children, as well as onboard preventive and control measures. The main infections occurring on board may be vaccine-preventable diseases (e.g. influenza, measles and varicella), respiratory infections (e.g. common cold, pharyngitis and middle-ear infection), diarrhoeal disease (e.g. rotavirus, norovirus and hepatitis A) and parasitic diseases (e.g. pediculosis or disease caused by hookworm). Crew members designated to onboard child-care play key roles in the prevention, surveillance and control of communicable diseases to and from children.

Background information

Types of child care that may be offered on board passenger ships include:

- care of infants and preschool children who are not toilet-trained and need supervision by staff, including changing diapers;
- care of children who are toilet-trained;
- provision and supervision of public playgrounds.

International standards and recommendations

None

Main risks

Ill-designed child-care facilities, lack of training for carers and inappropriate prevention and control procedures may pose a risk to all crew and passengers on board. General public health may also be at risk when children return to their communities.

Critical areas for controlling risks are:

- size, ventilation and lighting of child-care facilities;
- materials and cleanliness of surfaces of furniture, carpets and toys;
- diaper changing, hand-washing facilities and toilets;
- food preparation areas;
- training of crew in sanitation procedures, with an emphasis on hand washing;
- immunization of children and crew;
- communication procedures concerning notification of disease;
- isolation measures, including exclusion of sick children and crew from child-care facilities;
- methods to manage symptomatic passengers (such as isolation in cabin or departure from ship) to improve compliance with control measures.

Document review

Required documents are:

- written procedures and policies on cleaning, maintenance and waste management;
- written guidance on control measures if symptoms of infection occur in children; guidelines will include handling of body fluids, record keeping, notification of disease, communication, outbreak management and exclusion policies in case of illness;
- vaccination list of child-care staff.

References

International conventions

ILO, Maritime Labour Convention 2006.

Scientific literature

Carling PC, Bruno-Murtha LA, Griffiths JK (2009). Cruise ship environmental hygiene and the risk of norovirus infection outbreaks: an objective assessment of 56 vessels over 3 years. *Clinical Infectious Diseases*, 49:1312–1317.

Chimonas MA et al. (2008). Passenger behaviors associated with norovirus infection on board a cruise ship—Alaska, May to June 2004. *Journal of Travel Medicine*, 15:177–183.

Cliver D (2009). Control of viral contamination of food and environment. *Food and Environmental Virology*, 1:3–9.

Cramer EH, Gu DX, Durbin RE (2003). Diarrheal disease on cruise ships, 1990–2000: the impact of environmental health programs. *American Journal of Preventive Medicine*, 24:227–233.

McCutcheon H, Fitzgerald M (2001). The public health problem of acute respiratory illness in childcare. *Journal of Clinical Nursing*, 10(3):305–310.

Code of Areas	Inspection results: evidence found, sample results, documents reviewed	Control measures and corrective actions	Required	Recommended
	4.1 General design of child-care facility			
4.1.1 ❑	Child-care facilities not appropriate in size and location.	Provide child care in a space of appropriate size and location.		❑
4.1.2 ❑	Child-care facilities not well lit and not well ventilated.	Install sufficient lighting and/or ventilation.		❑
4.1.3 ❑	Presence of disease vectors such as insects. Presence of other sources of contamination. Facilities not clean.	Clean, disinfect and/or apply insecticide as appropriate.		❑
4.1.4 ❑	Potable water and hand-washing facilities not available and/or not appropriate for use by children.	Install appropriate washbasin with hot and cold potable water.		❑
4.1.5 ❑	Paper towels or hand-drying device, liquid soap, waste receptacle, toilet brush or toilet paper missing.	Equip room with all required materials.		❑
4.1.6 ❑	No separate toilets for staff and children, or toilets not appropriate in size for children; toilets dirty or do not flush properly.	Provide separate toilet facilities for children and staff. Clean toilets. Repair toilet flushing system as appropriate.		❑
4.1.7 ❑	Surfaces not smooth and durable. Carpets, toys and furniture not cleanable.	Equip room with appropriate material.		❑
	4.2 Diaper-changing facilities			
4.2.1 ❑	An area specifically set aside for diaper changing not provided within the facility.	Designate appropriate diaper-changing area.		❑
4.2.2 ❑	Diaper-changing area not adequately equipped. Hand-washing station, cleaning wipes, detergent, disinfectant and waste receptacle not available.	Equip diaper-changing area with appropriate material.		❑
	4.3 Training of staff			
4.3.1 ❑	Crew members designated for child care not trained in sanitary procedures or the symptoms and basic control of disease.	Train child-care staff in sanitary procedures and the symptoms and basic control of disease. Document training.		❑

Code of Areas	Inspection results: evidence found, sample results, documents reviewed	Control measures and corrective actions	Required	Recommended
4.4 Cleaning and disinfection				
4.4.1 ❑	Basic cleaning plan not maintained or carrying out the cleaning plan not documented.	Provide, observe and maintain cleaning plan.		❑
4.4.2 ❑	Written procedures not available for cleaning and disinfection of hands and materials in case of contact with blood, vomit or excrement.	Provide, observe and maintain cleaning plan.		❑
4.4.3 ❑	Written procedures for removal of waste not available.	Provide, observe and maintain waste management plan.		❑
4.4.4 ❑	Evidence of disease vectors and/or reservoirs found.	Disinfect, derat and apply insecticides, as appropriate.	❑	
4.5 Operation manuals				
4.5.1 ❑	Written guidance on control measures not available for when children show symptoms of common infections. Guidance to include control measures such as: •handling of body fluids; •record keeping; •notification and communication; •outbreak management and exclusion policies.	Provide written procedures and policies for responding to common childhood infections.		❑
4.6 Vaccinations				
4.6.1 ❑	Vaccination list of child-care crew members not present.	Provide updated vaccination list of crew.		❑

Area 5 Medical facilities

Introduction

Medical facilities are important for the onboard surveillance and control of disease. However, they also contribute to the occurrence of public health risks, as unsanitary conditions within medical facilities can cause the spread of communicable diseases. Sick passengers may pose a public health risk on board and ashore.

Therefore, crew members designated to provide onboard medical care play a key role in the prevention, surveillance and control of communicable diseases. Prerequisites for the control of public health risks on board include training of dedicated staff, appropriate operational manuals and protocols, facilities for diagnosis and treatment, and timely notification to the competent authority.

Smaller ships may not have the capacity to fulfil all measures for surveillance, prevention and control in the same way as larger ships with a physician on board.

International standards and recommendations

ILO, Maritime Labour Convention 2006

Regulation 4.1, Medical care on board ship and ashore: Standard A 4.1 stipulates that all ships carry a medicine chest, medical equipment and a medical guide. The national requirements shall take into account the type of ship, the number of persons on board and the nature, destination and duration of voyages and relevant national and international recommended medical standards. With ships carrying 100 or more persons and ordinarily engaged in international voyages of more than 3 days duration, a medical doctor who provides medical care shall be carried.

Guidelines B 4.1 and 2 outline the requirements to properly maintain and inspect the medicine chest by a designated crew member. Requirements include ensuring standardized medical training of designated seafarers, providing an up-to-date list of radio stations through which medical advice can be obtained, and carrying an appropriate medical report form.

ILO, IMO, WHO, International medical guide for ships, 3rd edition, 2008 (IMGS)

The IMGS is noted as a source of information in the non-statutory part of the Maritime Labour Convention 2006. It is a medical text that includes recommendations for the prevention, diagnosis, treatment and epidemic control of communicable diseases, including guidance on disinfection and removal of insects.

Chapter 33, Recommendations for the ship's medicine chest and equipment, includes specifications for the list of medicines and their storage, including record-keeping. Recommendations are given for antivirals, antimalaria medication, antibiotics, antipyretics, medication against diarrhoea, disinfectants for skin and wounds, personal protective equipment, thermometers and other items for the control of communicable diseases.

IMO, Medical first aid guide for use in accidents involving dangerous goods (MFAG) 1982

Ships carrying dangerous goods are obligated to have additional medicines, specific antidotes and special equipment on board, as prescribed in the MFAG.

IMO, International Convention on Standards, Certification and Watchkeeping for Seafarers 1978, as amended 1995 (STCW 95)

This convention is an international standard concerning the mandatory training of seafarers.

Main risks

Medical facilities need to be designed, equipped and maintained so that person-to-person spread of disease is prevented. All ships subject to the regulations established by the IMO and the ILO are required to follow training standards and carry a medical chest. The MFAG names specific medications and equipment that are mandatory for ships carrying dangerous goods. Beyond these requirements, no formal international instruments specify the contents of the medical chest, the design of medical facilities or operational manuals and protocols. The IMGS includes a suggested list of medications and equipment to be carried. A number of national maritime authorities further specify the contents and design of medical facilities plus the training of crew designated to onboard medical care.

Areas and standards to minimize the risk of spread of disease that concern medical facilities are as follows:

1. Medical facilities designed for accommodation of ill crew and passengers

Facilities must:

- be easily accessible and separated from other activities, particularly from food-storage and food-handling areas, and from spaces for waste;
- facilitate private treatment of ill travellers;
- be clean, well ventilated and well lit;
- provide adequate space for isolation of ill travellers;
- be properly maintained with potable water, and toilet and hand-washing facilities;
- not be used for other purposes.

2. Crew members designated to work in medical facilities

Staff must:

- be trained in basic medical first aid in accordance with STCW 95;
- include credentialed medical staff (physician and nurses) for ships carrying more than 100 people, in accordance with the Maritime Labour Convention 2006;
- provide evidence of attendance at approved training courses observing SCTW 95 criteria;
- demonstrate knowledge and competence by observed practices, such as adequate record keeping.

3. Medications and medical equipment

Supplies must:

- include all medications, personal protective equipment, medical devices and disinfectants sufficient to diagnose, treat and control public health risks according to the ship's size, number of travellers and voyage pattern;
- meet the recommendations and requirements of the IMGS and MFAG as a minimum.

Medications must:

- be given to travellers and crew only by trained and authorized personnel;
- be accompanied by adequate dispensory records.

Medical equipment must:

- be in good operational and hygienic order and operated and maintained according to manufacturers' recommendations.

4. Medical treatment log

A well-organized, legible and up-to-date medical log must list cases of illness, passengers and crew concerned, and any medication dispensed. Log entries should list:

- first date of clinic visit; name, age and sex of patient;
- passenger or crew member designation;
- crew member position or job;
- cabin number;
- date and time of illness onset;
- symptoms;
- details of specimen collection or other action taken, if applicable.

5. Confidentiality of personal medical and health information

Personal medical and other health information concerning passengers and crew, maintained in the above records or otherwise, must be processed and maintained confidentially in accordance with applicable laws and regulations.

6. Operational manuals

Procedures to reduce onboard risks of diseases should:

- relate to the ship's size, number of travellers, mix of patients, voyage pattern and the type and size of medical facilities;
- give special attention to adequate surveillance on board passenger ships (e.g. gastrointestinal disease log) and operation of high-risk facilities, such as haemodialysis or intensive-care units;
- provide adequate policies and procedures on cleaning, sanitation, sharps disposal and waste management.

7. Communication infrastructure

Communication infrastructure and procedures should be in place to contact external support for emergency medical advice services (telemedical assistance service) in case of a health emergency and to alert the competent authority about public health risks on board.

Document review

Required documents are:

- up-to-date ship's log and/or medical logbook, including treatment list;
- crew member interviews if the medical log is not available during inspection or entries are inadequate; if written information is required, request Maritime Declaration of Health from the State Party;
- training and certification of staff assigned to medical care;
- lists of medicines, vaccines, disinfectants and insecticides;
- number of passengers, mix of patients (passenger ships only), medical equipment in place and procedures performed, depending on ship's voyage pattern and size;
- cleaning, sanitation, maintenance and waste policies and procedures;
- specific disease surveillance logs (e.g. gastrointestinal disease), where applicable;
- operational manuals for high-risk facilities and devices such as an intensive-care unit, blood transfusion facility, operating theatre or haemodialysis facility;
- specimens collected and results if disease occurs on board; if possible, international certificates of vaccination or prophylaxis.

References

International conventions

ILO, Maritime Labour Convention (2006).

ILO, IMO, WHO (2008). *International medical guide for ships*, 3rd edition (http://apps.who.int/bookorders/anglais/detart1.jsp?sesslan=1&codlan=1&codcol=15&codcch=3078).

IMO, *Medical first aid guide for use in accidents involving dangerous goods*.

IMO, International Convention on Standards, Certification and Watchkeeping for Seafarers 1978, as amended in 1995, chapter VI.

Scientific literature

Anonymous] (2002). Norovirus activity—United States, 2002. *Morbidity and Mortality Weekly Report*, 52:41–45.

[Anonymous] (2003). The healthy traveler: cruising past infection. *Johns Hopkins Medical Letter: Health After 50*, 15:6.

Brotherton JM et al. (2003). A large outbreak of influenza A and B on a cruise ship causing widespread morbidity. *Epidemiology and Infection*, 130:263–271.

Centers for Disease Control and Prevention (1999). *Preliminary guidelines for the prevention and control of influenza-like illness among passengers and crew members on cruise ships*. Atlanta, Centers for Disease Control and Prevention.

Cramer EH et al. (2006). Epidemiology of gastroenteritis on cruise ships, 2001–2004. *American Journal of Preventative Medicine*, 30:252–257.

Cramer EH, Gu DX, Durbin RE (2003). Diarrheal disease on cruise ships, 1990–2000: the impact of environmental health programs. *American Journal of Preventative Medicine*, 24:227–233.

Dahl E (2004). Dealing with gastrointestinal illness on a cruise ship—Part 1: Description of sanitation measures. Part 2: An isolation study. *International Maritime Health*, 55:19–29.

Dahl E (2005). Medical practice during a world cruise: A descriptive epidemiological study of injury and illness among passengers and crew. *International Maritime Health*, 56:115–128.

Dahl E (2006). Norovirus challenges aboard cruise ships. *International Maritime Health*, 57:230–234.

Enserink M (2006). Infectious diseases. Gastrointestinal virus strikes European cruise ships. *Science*, 313:747.

Ferson MJ, Ressler KA (2005). Bound for Sydney town: health surveillance on international cruise vessels visiting the Port of Sydney. *Medical Journal of Australia*, 182:391–394.

Herwaldt BL et al. (1994). Characterization of a variant strain of Norwalk virus from a food-borne outbreak of gastroenteritis on a cruise ship in Hawaii. *Journal of Clinical Microbiology*, 32:861–866.

O'Mahony M et al. (1986). An outbreak of gastroenteritis on a passenger cruise ship. *Journal of Hygiene (London)*, 97:229–236.

Peake DE, Gray CL, Ludwig MR, Hill CD (1999). Descriptive epidemiology of injury and illness among cruise ship passengers. *Annals of Emergency Medicine*, 33:67–72.

Rooney RM et al. (2004). A review of outbreaks of foodborne disease associated with passenger ships: evidence for risk management. *Public Health Reports*, 119:427–434.

Schlaich CC, Oldenburg M, Lamshoft MM (2009). Estimating the risk of communicable diseases aboard cargo ships. *Journal of Travel Medicine*, 16:402–406.

WHO (1988). *International medical guide for ships*, 2nd ed., including the ship's medicine chest. Geneva, WHO.

Widdowson MA et al. (2004). Outbreaks of acute gastroenteritis on cruise ships and on land: identification of a predominant circulating strain of norovirus—United States, 2002. *Journal of Infectious Diseases*, 190:27–36.

Wilson ME (1995). Travel and the emergence of infectious diseases. *Emerging Infectious Diseases*, 1:39–46.

Code of Areas	Inspection results: evidence found, sample results, documents reviewed	Control measures and corrective actions	Required	Recommended
		5.1 Construction		
5.1.1 ❑	Medical facilities used for non-medical purposes (e.g. living quarters or storage rooms).	Use facilities for medical purposes only if ships carry 15 or more passengers and crew and are engaged in a voyage of more than 3 days.	❑	
		Provide accommodation exclusively for medical use.		❑
5.1.2 ❑	Medical facilities not easily accessible and not separated from other activities, food-storage and food-handling areas and waste disposal areas; and/or not suitable for private treatment of ill crew and passengers.	Provide private treatment space in a dedicated and suitable location.	❑	
5.1.3 ❑	Medical facilities not well lit and not well ventilated.	Install sufficient lighting for proper medical practices; evaluate sanitary conditions and/or ventilation.	❑	
		5.2 Equipment		
5.2.1 ❑	Potable water and hand-washing facilities not available.	Install washbasin with hot and cold potable water.	❑	
5.2.2 ❑	Paper towels or hand-drying device, liquid soap, waste receptacle, toilet brush or toilet paper missing.	Equip room with all required materials.	❑	
5.2.3 ❑	Absence or inadequate sharps or biomedical collectors.	Equip room with United Nations–certified sharps or biomedical collectors (for specifications, see Area 7, Solid and medical waste).	❑	
		5.3 Medical chest		
5.3.1 ❑	Size, type and storage of medical chest inadequate to diagnose, treat and control public health risks on board.	Supply adequate medications and/or equipment according to IMGS (3rd ed.) or flag-state requirements and MFAG if dangerous goods are on board.	❑	
5.3.2 ❑	Evidence of medicines that have passed expiry dates.	Replace all medicines that have passed their expiry dates with fresh medicines.	❑	
5.3.3 ❑	Medicines not stored according to manufacturers' requirements (e.g. vaccines not stored in refrigerators).	Store medicines according to manufacturers' requirements.	❑	
5.3.4 ❑	Medicines not stored in an organized manner.	Implement storage management of medicines; organize by medicine types, identification codes, etc.		❑

Code of Areas	Inspection results: evidence found, sample results, documents reviewed	Control measures and corrective actions	Required	Recommended
		5.4 Cleaning and maintenance		
5.4.1 ❑	No evidence of procedures and policies on cleaning, sanitation, sharps disposal or waste management.	Provide written procedures and policies relating to the complexity of medical care on board.		❑
5.4.2 ❑	Evidence of disease vectors and/or reservoirs that harbour disease vectors.	Disinfect, derat and apply insecticides immediately.	❑	
5.4.3 ❑	Toilet dirty or not flushing properly.	Clean toilets; repair toilet flushing system.		❑
5.4.4 ❑	Medical equipment and devices not in good operational and hygienic condition and not operated and maintained according to manufacturers' recommendations.	Operate and maintain equipment and medical devices according to manufacturers' recommendations.		❑
		5.5 Training of crew members		
5.5.1 ❑	No crew members designated to medical care, medication dispensing and maintenance of medical facilities on board.	Designate crew members for medical care.	❑	
5.5.2 ❑	For ships carrying 100 or more persons and ordinarily engaged in international voyages of more than 3 days, no medical doctor present to provide medical care.	Designate medical doctor to provide care, if applicable.	❑	
5.5.3 ❑	Crew members designated to work in medical facilities are not trained in basic medical first aid. No evidence of attendance at approved training courses observing SCTW 95 criteria. Designated staff demonstrate lack of knowledge and competence by observed poor practices.	Designate crew members with appropriate training according to level of onboard care provided.	❑	
		5.6 Health information		
5.6.1 ❑	Medical log not available during inspections. Entries not legible or up to date.	Provide updated medical log listing cases of illness, passengers or crew concerned and medications dispensed.	❑	
5.6.2 ❑	No up-to-date medical guide available (according to flag-state regulation or IMGS).	Provide up-to-date medical guide	❑	

Code of Areas	Inspection results: evidence found, sample results, documents reviewed	Control measures and corrective actions	Required	Recommended
5.6.3 ❑	No operations manuals available for the prevention, surveillance and control of public health risks on board (applicable to passenger ships only).	Provide surveillance logs for diseases (e.g. gastrointestinal diseases) and operations manuals for all medical procedures that are available on board.		❑
5.6.4 ❑	No evidence of adequate operations manuals for high-risk facilities and procedures (applicable only if such facilities are present).	Provide written procedures and policies for operation of high-risk facilities, such as intensive-care units and haemodialysis facilities.		❑
5.7 Communication infrastructure				
5.7.1 ❑	No or inadequate communication infrastructure and procedures in place to contact telemedical assistance service and to alert the competent onboard authority on public health risks.	Equip medical facilities with communication infrastructure and procedures.		❑
5.7.2 ❑	List of radio stations for telemedical assistance not available or up to date.	Provide up-to-date list of radio stations.	❑	

Area 6 Swimming pools and spas

Introduction

A variety of infectious agents (viruses, bacteria and protozoa) are associated with the recreational use of water, and these can affect the skin, ears, eyes, gastrointestinal tract and respiratory tract.

Risk factors for acquiring an infectious disease from the recreational use of water include:

- the presence of infectious agents;
- suitable conditions for the growth of infectious agents, e.g. temperature of 30–40 °C, a source of nutrients (organic matter from people bathing);
- a way of exposing crew and passengers to the infectious agents (e.g. Legionella bacteria in the aerosol created by water agitated in spa pools);
- the presence of people who could be exposed to the infectious agents (e.g. people passing near a spa pool).

Infectious agents can easily be introduced to pools and spa pools through people bathing, from dirt entering the pool or from the water source.

Spa pools are smaller than swimming pools and have a much higher ratio of bathers to water volume. Thus, the concentration of organic matter in spa pools is often far higher than in swimming pools. Water disinfection is, therefore, a key control measure, but the raised temperature and high organic content of spa pool water can make it difficult to maintain effective disinfection.

Non-microbiological hazards are also associated with the recreational use of water. These include accidental drowning, slipping, tripping and entrapment, as well as chemical, thermal and manual handling injuries.

Identification and assessment of the risks should be carried out to enable the ship operator to decide on measures to prevent or control exposure to infectious agents and other non-microbiological hazards. The ship operator is responsible for:

- assessing the risks associated with operating the swimming pool or spa pool;
- preventing or controlling exposure to hazards associated with the swimming pool or spa pool;
- developing, maintaining and field testing public health measures to control exposure;
- training crew members to use the control measures correctly.

Swimming pools and spa pools must be safe and free from irritants, infectious agents and algae. Daily maintenance of swimming pools and spa pools must:

- remove suspended and colloidal matter and render the water clear, bright and colourless;
- remove organic matter;
- provide sufficient disinfectant to control the growth of infectious agents;
- maintain the pH of the water at an optimum for disinfection;
- maintain a comfortable temperature for bathing.

Water treatment involves two main steps:

- filtration to maintain a physically clean, clear and safe environment;
- chemical disinfection to prevent cross-infection between people bathing and prevent the growth of infectious agents within the water and on surfaces in the swimming pool or spa pool and associated water and air circulation systems.

Effective purification relies on powerful filtration in conjunction with continuous disinfection via a complete and reliable circulation system to collect and disinfect water. To minimize contamination of the pool with organic matter from people bathing, it is essential to advise them to use toilets and shower before using the pool.

International standards and recommendations

WHO (2006). Guidelines for safe recreational waters, volume 2—swimming pools and similar recreational water environments. Geneva, WHO.

Main risks

The main risks are:

- microbiological (viruses, bacteria and protozoa);
- non-microbiological (accidental drowning, slipping or tripping, entrapment, as well as chemical, thermal and manual handling injuries).

Document review

Required documents are:

- schematic plan for recreational water facilities, plant and systems;
- written scheme for controlling the risk from exposure to disease-causing microorganisms;
- pool installation, design and construction, maintenance and operation specifications;
- training records for crew responsible for control methods;
- monitoring records;
- test results (e.g. pH, residual chlorine and bromine levels, temperature, microbiological levels);
- regular cleaning procedures;
- emergency cleaning and disinfection procedures.

References

Guidelines and standards

WHO (2006). *Guidelines for safe recreational water environments, volume 2—swimming pools and similar recreational-water environments*. Geneva, WHO.

WHO (2007). Legionella *and the prevention of legionellosis*. Geneva, WHO.

Scientific literature

Beyrer K et al. (2007). Legionnaires' disease outbreak associated with a cruise liner, August 2003: epidemiological and microbiological findings. *Epidemiology and Infection*, 135:802–810.

Chimonas MA et al. (2008). Passenger behaviors associated with norovirus infection on board a cruise ship—Alaska, May to June 2004. *Journal of Travel Medicine*, 15:177–183.

Goutziana G et al. (2008). *Legionella* species colonization of water distribution systems, pools and air conditioning systems in cruise ships and ferries. *BMC Public Health*, 8:390.

Jernigan DB et al. (1996). Outbreak of Legionnaires' disease among cruise ship passengers exposed to a contaminated whirlpool spa. *Lancet*, 347(9000):494–499.

Kura F et al. (2006). Outbreak of Legionnaires' disease on a cruise ship linked to spa-bath filter stones contaminated with *Legionella pneumophila* serogroup 5. *Epidemiology and Infection*, 134:385–391.

Rowbotham TJ (1998). Legionellosis associated with ships: 1977 to 1997. *Communicable Disease and Public Health* 1:146–151.

Code of Areas	Inspection results: evidence found, sample results, documents reviewed	Control measures and corrective actions	Required	Recommended
		6.1 Management		
6.1.1 ❑	Management plan absent, and/or responsible crew members unable to demonstrate knowledge of and/or competency in any or all of the following: • correct operation of the pool systems; • checks to be carried out (and their frequency) to ensure that the scheme is effective; • precautions to control the risk of exposure to disease-causing microorganisms.	Produce and implement a management plan for controlling the risk of exposure to disease-causing microorganisms.	❑	
		6.2 Design and construction		
6.2.1 ❑	Materials or fittings support the growth of microorganisms or corrode easily.	Replace materials or fittings with corrosion-resistant materials that do not support the growth of microorganisms.		❑
6.2.2 ❑	Pipework not accessible for cleaning; balance tanks not accessible for cleaning and disinfection.	Ensure that pipework and balance tanks are accessible for cleaning and disinfection.	❑	
		6.3 Equipment		
6.3.1 ❑	Ultraviolet (UV) disinfection device installed but not properly maintained and/or turbidity of the water greater than 0.5 nephelometric turbidity units (NTU).	Maintain UV disinfection devices according to manufacturers' instructions.	❑	
		Control turbidity of the water so that it is less than 0.5 NTU.	❑	
		6.4 Operation, cleaning and maintenance		
6.4.1 ❑	Water treatment programme absent.	Produce and implement a water treatment programme to include the use of chemicals and biocides, where appropriate.		❑
6.4.2 ❑	Chemicals and biocides not used to control microbiological activity and/or automatic chemical dosing pumps and equipment not regularly calibrated.	Ensure that automatic chemical dosing pumps and equipment are well maintained and regularly calibrated.	❑	
6.4.3 ❑	Responsible crew members unable to demonstrate knowledge of, and/or competency in, the operation and maintenance of the pool systems.	Train responsible crew members in knowledge of, and/or competency in, the operation and maintenance of the pool systems. Assess knowledge and/or competency following training.		❑
6.4.4 ❑	Responsible crew members unable to demonstrate knowledge of correct manual chemical dosing procedures.	Train responsible employees in the knowledge of, and/or competency in, the application of the water treatment programme.	❑	

<table>
<tr><th>Code of Areas</th><th>Inspection results: evidence found, sample results, documents reviewed</th><th>Control measures and corrective actions</th><th>Required</th><th>Recommended</th></tr>
<tr><td rowspan="4">6.4.5
❑</td><td rowspan="4">Operational parameters out of acceptable range for spa pools. Acceptable values are as follows:
• Free chlorine should not exceed 3 mg/l in public and semi-public pools and 5 mg/l in hot tubs.
• Bromine should not exceed 4 mg/l in public and semi-public pools and 5 mg/l in hot tubs.
• pH range 7.2–7.8 for chlorine disinfectants.
• pH range 7.2–8.0 for bromine-based and other non-chlorine processes.
• Turbidity < 0.5 NTU.

Microbiological value values out of acceptable range (see WHO Guidelines for safe recreational water environments, volume 2, Table 5.3 Recommended routine sampling frequencies and operational guidelines for microbial testing during normal operation)

Responsible employees are unable to demonstrate knowledge of and/or competency in the safe operating limits of the defined parameters.</td><td>Close pools and check pH calibration and value, then rectify faults and recheck pH. If pH is still out of limits, empty pool and refill with clean water to reach pH 7.2 and add appropriate level of disinfectant.</td><td></td><td>❑</td></tr>
<tr><td>Check dosing units and calibration are working properly (i.e. they contain adequate disinfectant, the flow rate is appropriate and there are no air locks or blockages in pipework).</td><td>❑</td><td></td></tr>
<tr><td>Train responsible crew members in knowledge of, and/or competency in, the monitoring of safe operating limits. Assess knowledge and/or competency following training.</td><td></td><td>❑</td></tr>
<tr><td>Take microbiological samples of the water and analyse at least heterotrophic plate count, Escherichia coli, Pseudomonas aeruginosa, Legionella spp.</td><td>❑</td><td></td></tr>
<tr><td>6.4.6
❑</td><td>Responsible crew unable to demonstrate knowledge of, and/ or competency to carry out, corrective action for agreed out-of-limit situations.</td><td>Train responsible crew in knowledge of, and/ or competency in carrying out, corrective action for agreed out-of-limit situations. Assess knowledge and/or competency following training.</td><td>❑</td><td></td></tr>
<tr><td>6.4.7
❑</td><td>Responsible crew unable to demonstrate that checks are made on:
• cleanliness of the water in the system;
• backwash of sand filters;
• cleanliness of water line, overflow channels, grills and pool surrounds.</td><td>Train responsible crew in knowledge of the maintenance procedures.</td><td>❑</td><td></td></tr>
<tr><td>6.4.8
❑</td><td>Responsible crew unable to demonstrate knowledge of the determined control levels of biocides and the rate of release or rate of addition of biocide.</td><td>Train responsible crew in knowledge of the determined control levels of biocides and the rate of release or rate of addition of biocide.</td><td>❑</td><td></td></tr>
</table>

Code of Areas	Inspection results: evidence found, sample results, documents reviewed	Control measures and corrective actions	Required	Recommended
6.4.9 ❑	Microbiological tests for indicator organisms not carried out.	Test for indicator microorganisms regularly.	❑	
6.4.10 ❑	Seawater flow-through pools operated in port without suitable water treatment.	Close seawater flow-through pools for use whenever ship is in port or in other water bodies that may harbour potential contamination.	❑	
6.4.11 ❑	Pool surrounds, overflow channels, exposed pipework, filters and fittings visibly dirty or greasy.	Clean dirty components immediately.	❑	
		Produce and implement a procedure for regular cleaning of the pools.		❑
6.4.12 ❑	Responsible crew unable to demonstrate knowledge of, and/ or competency in, the regular cleaning procedures of the pool systems.	Train responsible crew in knowledge of the regular cleaning procedures of the pool systems. Assess knowledge and/or competency following training.	❑	
6.4.13 ❑	Surfaces in sauna appear dirty and not well maintained.	Clean and disinfect all surfaces that come in contact with people, to avoid spread of diseases (e.g. skin diseases).		❑
6.4.14 ❑	Responsible crew unable to demonstrate that good hygiene information is available to pool users.	Provide good hygiene information for pool users, e.g. signage showing "Use toilets and showers" and "Avoid submerging your head in spa pools".	❑	
6.4.15 ❑	Evidence of disease vectors found.	Apply disease vector control measures and disinfect.	❑	
6.5 Emergency procedures				
6.5.1 ❑	No faecal accident response procedure defined and/or responsible crew are unable to demonstrate knowledge of, and/ or competency in, emergency cleaning and disinfection procedures.	Produce and implement a procedure for emergency cleaning and disinfection. Train responsible employees in knowledge of emergency cleaning and disinfection procedures. Assess knowledge and/or competency following training.		❑

Area 7 Solid and medical waste

Introduction

Depending on the type and route of a ship, large amounts of waste are produced on board. This waste can be separated into food waste, paper and cardboard, cans and tins, glass, plastics, oily materials and potentially infectious medical waste.

The international definition of garbage is all kinds of food, domestic and operational waste, excluding fresh fish and parts of fish, generated during normal operation of the ship, as defined in Annex V of the IMO International Convention for the Prevention of Pollution from Ships (MARPOL 73/78).

While MARPOL regulations are targeted to environmental protection, unsafe management and disposal of ship waste can also lead to adverse health consequences. MARPOL Annex V details the proper retention, selective collection, storage and disposal of wastes on board, on shore and overboard (where shore areas will not be affected). MARPOL Annex V includes measures to prevent the creation of health hazards.

It is necessary to meet international standards and recommendations to help avoid pollution of the seas and creation of individual and public health risks.

The recommendations in this checklist follow the physical waste stream of production–transport–processing–storage–disposal.

International standards and recommendations

IMO, International Convention for the Prevention of Pollution from Ships 1973, as modified by the protocol of 1978 relating thereto (MARPOL 73/78) as amended.
Annex V: Prevention of pollution by garbage from ships.

Garbage type	Outside special areas	In special areas
Plastics, including synthetic ropes, fishing nets and plastic bags	Disposal prohibited	Disposal prohibited
Floating dunnage, lining and packing material	Disposal prohibited less than 25 nautical miles from nearest land	Disposal prohibited
Paper, rags, glass, metal, bottles, crockery and similar refuse	Disposal prohibited less than 12 miles from nearest land	Disposal prohibited
Paper, rags, glass, etc., comminuted or ground[a]	Disposal prohibited less than 3 miles from nearest land	Disposal prohibited
Food waste, comminuted or ground[a]	Disposal prohibited less than 3 miles from nearest land	Disposal prohibited less than 12 miles from nearest land
Food waste, not comminuted or ground	Disposal prohibited less than 12 miles from nearest land	Disposal prohibited less than 12 miles from nearest land
Mixed refuse	Varies by component[b]	Varies by component[b]

[a] Must pass through a screen with a mesh size no larger than 25 mm.
[b] When substances with different disposal or discharge requirements are mixed, the more stringent disposal requirement applies.

Codex Alimentarius Commission (2003). CAC/RCP1-1969 (Rev.4-2003) Recommended international code of practice—general principles of food hygiene; incorporates hazard analysis and critical control point (HACCP) system and guidelines for its application.
WHO (2004). Rolling revision of the Guidelines for drinking-water quality and Guide to ship sanitation (Draft), 10/2004. Geneva, WHO.
WHO (1999). Safe management of wastes from health-care activities. Geneva, WHO.
IMO (2000). Guidelines for ensuring the adequacy of port waste reception facilities. London, IMO.
IMO Resolution MEPC.70(38): Guidelines for the development of garbage management plans. London, IMO.
IMO Resolution MEPC.76(40): Standard specification for shipboard incinerators. London, IMO.
IMO Convention on Facilitation of International Maritime Traffic 1965, as amended, 2006 edition. Annex 5: Certificates and documents required to be carried on board ships.

Main risks

Food wastes attract disease vectors, including rodents, flies and cockroaches. All waste can contain physical, hazardous microbial or chemical agents; for example, sharp objects such as needles may harbour infectious agents. Harmful chemicals can be deposited in waste and pose a risk for waste-handling staff.

Humans can become exposed directly, both on board and at port, by contact with waste that is not managed in a safe manner. Exposure can also occur through the environmental transfer of disease-causing organisms or harmful substances that have not been disposed of safely. However, waste can be managed and disposed of in ways that prevent harm. Procedures to facilitate the safe processing, storage and discharge of garbage should be implemented in a garbage management plan.

Document review

Required documents:

- a garbage management plan for every ship of 400 tons gross tonnage and above, and every ship certified to carry 15 persons or more; this document should contain all information requested in the Marine Environment Protection Committee Guidelines for the development of garbage management plans;
- a garbage record book for every ship of 400 tons gross tonnage and above, and every ship certified to carry 15 persons or more; this document should contain information on amounts of different waste types produced on board, plus information including discharge and incineration processes;
- International safety management manual;
- maintenance instructions for waste processing units (e.g. incinerator);
- construction plans of sewage system to check drains in waste areas.

References

International conventions

IMO, Maritime Labour Convention 2006.

Guidelines and standards

WHO (1999). *Safe management of wastes from health-care activities.* Geneva, WHO.
WHO (2011). *Guide to ship sanitation.* Geneva, WHO.

Code of Areas	Inspection results: evidence found, sample results, documents reviewed	Control measures and corrective actions	Required	Recommended
7.1 Garbage record book Required for all ships >400 tons gross tonnage or with ≥15 persons on board				
7.1.1 ❑	No garbage record book available, or garbage record book does not contain all disposal and incineration operations.	All garbage record books must be available for a minimum of 2 years. Notify Port State Control.	❑	
	The date, time, position of ship, description of the garbage and estimated amount incinerated or discharged are not logged and/or signed.	Garbage record books must be up to date. Provide missing information to Port State Control and competent authority.	❑	
7.2 Garbage management plan Required for all ships of >400 tons gross tonnage or with ≥15 persons on board				
7.2.1 ❑	No garbage management plan available, or not all procedures for collecting, storing, processing and disposing of garbage are covered in the plan.	Develop garbage management plan according to IMO guidelines.	❑	
		Implement all procedures for collecting, storing, processing and disposing of garbage in the plan.	❑	
		Nominate a designated person to be in charge of carrying out the garbage management plan.	❑	
		Translate garbage management plan into the working language of the crew.	❑	
7.3 Management				
7.3.1 ❑	Lack of training materials and/or evidence of lack of knowledge about garbage procedures and discharge regulations.	Train crew in procedures and regulations relating to garbage collecting, processing, sorting and disposal.		❑
		Support crew with training materials about garbage separation, processing, storage and discharge.		❑
		Display signage in English, French or Spanish language, notifying passengers and crew of the disposal requirements according to MARPOL Annex V.	❑	
7.4 Places of waste production				
7.4.1 ❑	Waste containers for food refuse: • are not available; • are dirty; • are not tightly covered; • are not watertight; • emit a strong odour; • attract rodents or other vermin. Places for food waste include galley, pantry and restaurants.	Install waste container that is watertight, nonabsorbent and easily cleanable; can be disinfected; and has a tightly fitting cover.	❑	
		Scrub, wash and disinfect containers thoroughly after each emptying.	❑	
		Control pests in the area.	❑	

Code of Areas	Inspection results: evidence found, sample results, documents reviewed	Control measures and corrective actions	Required	Recommended
7.4.2 ❑	Garbage containers for other garbage types are: • not watertight; • absorbent; • difficult to clean; • not equipped with tight covers.	Equip area with appropriate containers.	❑	
		Define an adequate waste container storage area.		❑
7.4.3 ❑	Existing containers are: • dirty; • broken; • attracting rodents or vermin; • places where disease vectors feed or breed, either inside or outside waste containers.	Clean, disinsect and disinfect dirty containers at an area away from any food areas.	❑	
		Clean, disinsect and disinfect affected area.	❑	
		Replace broken containers.	❑	
7.4.4 ❑	Waste containers not tightly covered between operations (e.g. opened containers may be necessary during food operations).	Cover waste containers in food preparation or serving areas during operations whenever possible.		❑
		Supply food areas with waste containers that can be opened without use of hands (e.g. with a foot pedal).		❑
7.4.5 ❑	Grease separated from galley wastes not handled correctly.	Install a grease interceptor between galley wastewater drains and wastewater system.	❑	
		Collect and dispose of grease in a legal way (e.g. authorized port waste reception facility, incineration or overboard discharge on the high seas).	❑	
		Clean grease trap.		❑
7.4.6 ❑	Evidence of vectors and/or reservoirs found.	Disinfect the reservoirs and apply vector control measures.	❑	
		7.5 Medical waste		
7.5.1 ❑	Accumulation of medical wastes at the point of production (e.g. medical facility).	Remove correctly packaged waste from the point of production to a dedicated secure storage place.	❑	
		Dispose of medical waste as soon as possible in an appropriate land-based facility.	❑	
		Reduce storage time as much as possible.		❑
		Include medical waste handling in the garbage management plan.	❑	
		Store potentially infectious waste in yellow plastic bags or containers that are labelled with the words "HIGHLY INFECTIOUS" and with the international infectious substances (biohazard) symbol.	❑	

Code of Areas	Inspection results: evidence found, sample results, documents reviewed	Control measures and corrective actions	Required	Recommended
7.5.2 ❑	Medical waste not disposed of in coloured and labelled plastic bags or containers, or not separately stored. Hazardous health-care waste not separated from non-hazardous waste.	Store non-infectious waste from health-care facilities in black plastic bags.		❑
		Store medical wastes separately from other wastes at a dedicated place.	❑	
		Supply fresh collection bags or containers.		❑
		Provide appropriate containers or bag holders in dispensary or medical facility.		❑
		Post waste separation and identification instructions at each waste collecting point.		❑
		Remove containers and bags when they are three-quarters full.		❑
		Close waste bags tightly (e.g. with cable straps).		❑
7.5.3 ❑	Liquid medical wastes not discharged into the sewage (black-water) system; dispensary or medical facility drains not connected to the black-water system.	Connect piping for medical liquid waste and wastewater from medical areas, including bathtubs, showers and hand-wash basins, to the sewage system.	❑	
7.5.4 ❑	Sharps waste (e.g. needles, blades) not stored in appropriate containers; waste other than sharps found in the container; container or containers are full.	Dispose of sharps waste into suitable plastic containers.	❑	
		Provide sharps containers that are made of metal or puncture-proof plastic, fitted with covers, rigid, impermeable and tamper proof. Containers need to be yellow, and labelled with the word "SHARPS" and the international infectious substance (biohazard) symbol.	❑	
		Provide appropriate container holders to avoid injuries if ship is rolling.		❑
		Dispose of all containers that are three-quarters full into a labelled, yellow infectious medical waste bag before removal from the dispensary.	❑	
		Do not dispose of any waste other than sharps in the sharps container.		❑
7.5.5 ❑	Pharmaceutical waste not stored or disposed of correctly.	Store pharmaceutical waste (e.g. out-of-date medicines) ideally in brown plastic bags and return it to a land-based disposal facility.		❑
		Prohibit incineration at low temperatures or discharge of pharmaceuticals into the sewage system.		❑
7.5.6 ❑	No designated secure storage place for safe storage and/or treatment of medical wastes.	Designate a storage place for medical waste and secure it against unauthorized access.	❑	
7.5.7 ❑	Plastics or wet materials found prepared to be incinerated.	Allow only paper- or cloth-based material to be incinerated.	❑	

Code of Areas	Inspection results: evidence found, sample results, documents reviewed	Control measures and corrective actions	Required	Recommended
7.5.8 ❑	Crew members handling potentially infectious health-care wastes not vaccinated against hepatitis B.	Vaccinate crew members in charge of handling these waste types against hepatitis B.		❑
7.5.9 ❑	Evidence of vectors and/or reservoirs found.	Disinfect, derat and apply insecticides immediately.	❑	
		7.6 Hazardous chemical waste		
7.6.1 ❑	No designated space for storage of hazardous chemical waste; area not secure against unauthorized access; area dirty, insufficiently lit or insufficiently ventilated.	Designate a space for storage of hazardous waste.	❑	
		Improve lighting in storage area.		❑
		Improve ventilation in storage area.		❑
		Clean storage place.	❑	
		Secure storage area against unauthorized access.	❑	
7.6.2 ❑	Hazardous chemical wastes of different composition accumulating or not stored separately.	Store different chemicals separately to avoid chemical reactions.	❑	
		Dispose of waste to approved organizations or agencies authorized to manage hazardous waste.	❑	
		Obtain information about suitable waste reception facilities in ports to minimize accumulation of hazardous waste.		❑
7.6.3 ❑	Evidence of vectors and/or reservoirs found.	Disinfect, derat and apply insecticides immediately.	❑	
		7.7 Transport		
7.7.1 ❑	Interiors of garbage lifts and chutes or other waste transport systems: • are not adequately constructed; • show dirt or grease accumulation; • have a strong odour; • are damaged or corroded.	Clean and disinfect waste transport facilities.	❑	
		Install removable, cleanable, non-absorbent and non-corroding cover with suitable integral cove of at least 10 mm along all sides for bottom of lifts.		❑
		Improve construction to allow waste transport systems to be easy to clean and disinfect.		❑
		Replace interiors of garbage lifts and chutes with stainless steel.		❑
		Provide garbage chutes with automatic cleaning systems.		❑
		Clean and disinfect all chutes and lifts regularly.	❑	
7.7.2 ❑	No drain provided at the bottom of lift shafts; drain not connected to the sewage system; or bottom of lift shaft dirty.	Install drains at bottom of lift shafts.		❑
		Connect drains to sewage system.		❑
		Clean and disinfect bottom of lift shaft.		❑
7.7.3 ❑	Evidence of vectors and/or reservoirs found.	Disinfect, derat and apply insecticides immediately.	❑	

Code of Areas	Inspection results: evidence found, sample results, documents reviewed	Control measures and corrective actions	Required	Recommended
		7.8 Garbage processing		
7.8.1 ❑	Sorting tables not constructed from impervious, non-absorbent material (preferably stainless steel), do not have coved corners and rounded edges, or are broken or dirty.	Clean and disinfect sorting tables carefully after each use.		❑
		Install sorting tables made of suitable material (preferably stainless steel), with coved corners and rounded edges. If deck coaming is provided, it should be at least 8 cm high and coved. Tables should be drained to the sewage system.		❑
7.8.2 ❑	Room where garbage is processed does not meet the same criteria as garbage storage rooms.	Provide garbage processing rooms with same equipment as garbage storage rooms, including ventilation, lighting, potable water hose and drains.		❑
7.8.3 ❑	No hand-washing facilities available close to waste processing areas. Hand-washing facilities not adequately equipped.	Provide hand-washing facilities with running hot and cold potable water.		❑
		Equip hand-washing facility with disposable towels, liquid soap, hand disinfection liquid, waste receptacle and signage showing "Wash and disinfect your hands".		❑
		Install a hose connection and sufficient drains to avoid pooling of water.		❑
7.8.4 ❑	No personal protective equipment (PPE) available; PPE not in good operational condition; and/or crew members show no competency in the use of PPE.	Equip staff in charge of waste handling with safety goggles or face shield, face mask, rubber gloves, working gloves, safety boots or shoes, and a protective suit. Train the staff.	❑	
7.8.5 ❑	Pieces of comminuted garbage that are collected for overboard disposal are too large.	Comminute garbage until it passes a mesh size of 25 mm before disposal.	❑	
7.8.6 ❑	Commuter and/or compactors in dirty condition, emit a strong smell or attract rodents and other vermin.	Clean, disinfect and derat garbage processing facilities.	❑	
7.8.7 ❑	Place used to clean garbage containers promotes cross-contamination and/or is in poor sanitary condition.	Designate a place to clean garbage containers far away from any food areas (e.g. in properly equipped and maintained garbage storage room).		❑
7.8.8 ❑	Evidence of vectors and/or reservoirs found.	Disinfect, derat and apply insecticides immediately.	❑	

Code of Areas	Inspection results: evidence found, sample results, documents reviewed	Control measures and corrective actions	Required	Recommended
		7.9 Storage		
7.9.1 ❑	Storage place does not meet the following requirements: • adequate size; • protected from the sun; • inaccessible to animals, insects and birds; • easy to clean and disinfect; • durable, non-absorbent, hard floor; • drainage to sewage system; • water supply for cleaning purposes; • water hose for cleaning; • easily accessible for staff in charge; • secured against unauthorized access; • good lighting (220 lux) and ventilation; • not situated in the proximity of fresh food stores or food preparation areas	Ensure garbage rooms are large enough to hold unprocessed waste for longest period expected between offloading of waste.	❑	
		Develop a cleaning schedule for regular cleaning and disinfection.		❑
		Install sun protection or change location of storage area to avoid heat from sun and other sources.	❑	
		Protect storage room from intrusion of animals and insects.	❑	
		Tightly cover containers stored on deck.	❑	
		Install drainage and connect drainage to sewage system.	❑	
		Provide running water and hose for cleaning purposes.		❑
		Secure garbage room or containers against unauthorized access.	❑	
		Improve ventilation and lighting.		❑
		Ensure garbage area is distant from food area.		❑
7.9.2 ❑	No hand-washing facilities available close to waste holding areas. Hand-washing facilities not adequately equipped.	Provide hand-washing facilities with running hot and cold potable water.		❑
		Equip hand-washing facilities with disposable towels, liquid soap, hand disinfection liquid, waste receptacle and signage stating "Wash and disinfect your hands".		❑
		Install a hose connection and sufficient drains to avoid pooling of water.		❑
7.9.3 ❑	No locker for cleaning materials available; broken or dirty equipment.	Provide storage locker with cleaning utensils away from food.		❑
		Provide proper cleaning equipment.	❑	
7.9.4 ❑	No adequate supply of cleaning equipment, PPE and waste bags or containers located close to the storage area.	Supply cleaning equipment.		❑
		Supply PPE, including safety goggles or face shield, rubber work gloves, face mask, safety shoes or boots, and protective suit.		❑
		Supply suitable waste bags and/or containers close to storage area.		❑
7.9.5 ❑	Inadequate supply of waste containers. No separation of garbage types or not enough suitable receptacles	Supply containers of adequate capacity for paper, plastics, tins, food waste and dry waste.	❑	
		Label waste containers according to content.	❑	
		Store different garbage types separately.	❑	
		Store dry and food wastes in tightly covered containers protected against the weather and from intrusion of rodents and other vermin.	❑	

Code of Areas	Inspection results: evidence found, sample results, documents reviewed	Control measures and corrective actions	Required	Recommended
7.9.6 ❑	Containers dirty or attracting disease vectors.	Thoroughly clean, disinfect and apply insecticides to the containers after each emptying if necessary.		❑
7.9.7 ❑	No refrigerated space available where necessary for wet refuse.	Provide a sealed refrigerated space for storing wet garbage; the space needs to meet the same criteria used for cold food storage.		❑
7.9.8 ❑	Garbage room full of garbage.	Dispose of garbage at port reception facility.		❑
		Gather information about suitable port waste reception facilities in the next ports.		❑
7.9.9 ❑	Vectors or reservoirs found.	Disinfect, derat and apply insecticides immediately.	❑	
		7.10 Incinerator		
7.10.1 ❑	Ash, plastic materials or other substances that may contain heavy metals or other poisonous substances discharged to the sea.	Train crew in correct disposal of incinerator ash.		❑
		Inform competent authority for MARPOL violations.	❑	
7.10.2 ❑	Evidence of leakage of gases and/or particles from combustion chamber.	Check exhaust system and gas-tightness of incinerator plant.		❑
7.10.3 ❑	Incinerator has no prominent signage to warn against unauthorized opening of doors during operation and against overloading incinerator with garbage.	Install appropriate signage stating "Do not open during operation" and "Do not overload incinerator".		❑
7.10.4 ❑	Incinerator full of ash or slag.	Clean combustion chamber and dispose of ash or slag at a port reception facility.		❑
7.10.5 ❑	Incinerator room is dirty or garbage is accumulating.	Clean incinerator room.	❑	
		Store garbage in an appropriate storage room.		❑
7.10.6 ❑	Vectors or reservoirs found.	Disinfect, derat and apply insecticides immediately.	❑	
		7.11 Discharge		
7.11.1 ❑	Garbage has been disposed of overboard in a special area, or any other evidence of prohibited waste deposit into water.	Inform competent authority (e.g. Port State Control).	❑	
		Capture and retain waste on board.		❑
7.11.2 ❑	Master or crew not familiar with procedures for managing garbage on board.	Display signage in English, French or Spanish, notifying passengers and crew of garbage disposal requirements in accordance with MARPOL Annex V.		❑

Area 8 Engine room

Introduction

The engine room and nearby compartments can contain hazardous microbial, chemical and physical agents. Infectious agents and harmful chemicals can be transferred from the engine room into waste through connections to black water (as defined in the glossary), grey water (as defined in the glossary), ballast water, effluent from oil–water separators, cooling water, boiler and steam generator blowdown, industrial wastewater and other hazardous waste.

People can become exposed directly through contact with onboard waste and facilities that are not managed in a safe manner. Exposure can also occur through the environmental transfer of disease-causing organisms or harmful substances due to unsafe environmental management, operational failure and lack of crew training.

International standards and recommendations

IMO, Guidelines for engine room layout, design and arrangement (MSC 68/Circ 834)

1. Chapter 6.3, Ergonomics:
6.3.7. The layout, design and arrangement of machinery and work areas in engine rooms should be such that the engine room can be conveniently cleaned.
6.3.9. A supply of consumables, such as light bulbs, flashlights, batteries, aural protection, protective goggles, disposable work clothes, gloves, rags, cups, logbooks, pens and pencils, should be maintained in the engine room for the use of personnel working in the engine room.

2. Chapter 6.4, Minimizing risk through layout and design:
6.4.6. Engine rooms should be provided with means for collecting and disposing of oil, paper, rags and other wastes and with supplies for cleaning to minimize the potential for fire and personnel injury.

IMO, International Convention for the Prevention of Pollution from Ships 1973, as modified by the protocol of 1978 relating thereto (MARPOL)

Annex VI sets limits on emissions of nitrogen oxides (NOx) from diesel engines.
Amendments to the annex of the protocol of 1997 to amend the International Convention for the Prevention of Pollution from Ships 1973, as modified by the protocol of 1978 relating thereto (revised MARPOL Annex VI): Progressive reductions in nitrogen oxide (NOx) emissions from marine engines were also agreed, with the most stringent controls on so-called "Tier III" engines, i.e. those installed on ships constructed on or after 1 January 2016, operating in emission control areas.

ILO Convention No. 133 on Accommodation of Crews 1970

1. Article 9: Ships of 1,600 tons or over shall be provided:
(b) a water closet and a wash basin having hot and cold running fresh water, within easy access of the machinery space if not fitted near the engine room control centre.

2. In ships of 1,600 tons or over, other than ships in which private sleeping rooms and private or semi-private bathrooms are provided for all engine department personnel, facilities for changing clothes shall be provided, which shall be:
(a) located outside the machinery space but with easy access to it; and
(b) fitted with individual clothes lockers as well as with tubs and/or shower baths and wash basin having hot and cold running fresh water.

Main risks

The main risks include contamination by vectors and effects of the engine-room environment on the occupational health of crew members, including:

- external exposure to oil and inhalation of oil vapour and mist created by poorly ventilated and located equipment;
- high temperature of enclosed control rooms with insufficient cooling facilities;
- insufficient lighting.

Document review

None applicable.

References

International conventions

IMO (1978). *Guidelines for engine room layout, design and arrangement* (MSC 68/Circ 834). London, IMO. Chapter 6.3, Ergonomics; Chapter 6.4, Minimizing risk through layout and design.

IMO, International Convention for the Prevention of Pollution from Ships 1973, as modified by the Protocol of 1978 relating thereto (MARPOL), Annex VI.

Code of Areas	Inspection results: evidence found, sample results, documents reviewed	Control measures and corrective actions	Required	Recommended
		8.1 Construction		
8.1.1 ❑	Construction design not favourable for cleaning.	Design and arrange machinery and work areas so that the engine room can be conveniently cleaned.		❑
		8.2 Management		
8.2.1 ❑	Evidence of vectors found.	Apply vector control measures and eliminate vector reservoir.	❑	
8.2.2 ❑	Ducts extend from the weather deck directly to the engine room without protection from vectors.	Vector-proof both ends of the service outlet of cold-air or hot-air systems serving more than one compartment.		❑
		8.3 Equipments and facilities All ships of 1,600 tons or more		
8.3.1 ❑	Hand-washing station not within easy access.	Install a washbasin with hot and cold potable water within easy access of machinery.	❑	
8.3.2 ❑	Washing station and clothes-changing room for engine department personnel absent.	Provide facilities for changing clothes that are: • located outside the machinery space but within easy access; • fitted with individual clothes lockers as well as showers or baths and washbasins with hot and cold running potable water.	❑	
		8.4 Ventilation		
8.4.1 ❑	Ventilation units are out of order.	Repair or replace ventilation units.		❑

Area 9 Potable water

Introduction

Clean drinking-water is essential for health; therefore, nearly every nation in the world has their own set of regulations to assure clean drinking-water for its population. Countries that have not defined their own drinking-water regulations often refer to the WHO Guidelines for drinking-water quality (GDWQ), Vol. 1, 3rd ed. Geneva: WHO.

Ships may be equipped with two or three different water systems: potable water, non-potable water used for other operational procedures and water for firefighting. Whenever practicable, only one water system should be installed to supply potable water for drinking, personal hygiene, culinary purposes, dishwashing, and hospital and laundry purposes. Non-potable water, if used on the ship, needs to be loaded and distributed through a completely different piping system, which should be colour coded according to existing international standards.

Definition of drinking-water

The terms "drinking-water" or "potable water" are used to define any water for human consumption. This includes not only water for drinking or cooking, but also water for brushing teeth, showers, washing hands, washing clothes and so on. Even on large, modern merchant vessels, showers and washbasins that have so-called fresh water are actually drawing it directly from desalination plants; therefore, it does not meet drinking-water quality criteria. Untreated "fresh water" may also harbour many health risks for the consumer and public health. Therefore, ships equipped with "freshwater systems" cannot use water that is unfit for any human consumption.

International standards and recommendations

ILO, Accommodation of Crews (Supplementary Provisions) Convention 1970 (No. 133)

The ILO Convention no. 133 has been ratified by a large number of Member States. It defines the minimum standards for crew accommodation on commercial ships above 1,000 gross tonnes. This convention states that people on board need to have permanent access to cold potable water. Additionally, showers or bathtubs, and washbasins must have running hot and cold "fresh water". The definition of this term is problematic, as described above.

The ILO Convention no. 133 will be included in the ILO Maritime Labour Convention 2006, which defines the same requirements, but is still in process of ratification.

WHO *Guidelines for drinking-water quality*

This document gives information about microbial, physical and chemical aspects of drinking-water quality, and is often used as a reference by different national drinking-water legislation.

International Organization of Standardization (ISO)

ISO has published several important international standards that describe technical aspects for safe potable water constructions.

Main risks

Improperly managed water on ships is an established route for infectious disease transmission. Furthermore, water may be a source of index cases of disease, which might then be transmitted via other routes. Most waterborne outbreaks involve ingestion of water that was contaminated with pathogens derived from human or animal excreta. Contamination is associated with spoiled bunkered water, cross-connections between potable and non-potable water, improper loading procedures, poor design and construction of potable water storage tanks, and inadequate disinfection. Space is often limited on board ships, and therefore potable water systems are likely to be physically close to excessive heat, or close to hazardous substances such as sewage or waste streams. Avoiding cross-contamination is one of the major challenges of keeping water safe on ships.

Bunkering is a high-risk procedure. Unsafe handling and using inappropriate materials (e.g. firefighting hoses) may lead to contamination. Hazardous water can be supplied from shore and, if the ship has no barrier systems, the water may contaminate the ship's potable water. Contamination of the shore (supply) system is also possible if there are no backflow preventers installed between the shore and the ship. It is crucial to know where risks exist, and it is necessary to implement good handling practice concerning potable water hoses and the whole bunkering procedure.

The GDWQ define the recommended minimum quality criteria of potable water. Some of the most common criteria are listed below. The parameters in bold can be used as on-site parameters to monitor a ship's water safety.

Appearance and colour
The appearance and taste of drinking-water should be acceptable to the consumer. Water should have no detectable odour. Ideally, drinking-water should have no visible colour.

pH
For effective disinfection with chlorine, the pH should be less than 8.0. The optimum pH depends on the water and the materials used in the potable water distribution system, but it is usually in the range 6.5–8.0 and can extend up to 9.5. The pH is important when checking water treatment efficiency and corrosive potential of mains and pipes in the drinking-water distribution system.

Temperature
Water temperature should always be either below 25 °C or above 50 °C. In temperatures of 25–50 °C, a high risk of bacterial growth (especially Legionella spp.) exists, and water safety testing should be performed.

Conductivity
Electrical conductivity is not discussed in the GDWQ. It is an important operational parameter to assess the efficacy of the desalinated water remineralisation process. Typical values (in µS/cm) for desalinated water are very low. A contamination of distillate or seawater can be easily detected because seawater has a high conductivity (e.g. 50 000 µS/cm).

Chlorine
Free chlorine and total chlorine should be measured during or after the disinfection treatment, or more often as required. Effective disinfection should have a free chlorine (Cl2) concentra-

tion of between 0.5 mg/l and 1.0 mg/l at the point of consumption. Different national standards in chlorination levels should be considered, because they can differ. For example, the United States Vessel Sanitation Program states that the acceptable minimum level of free chlorine at the distant point while water should be in consumption condition is 0.2 mg/l, whereas in European the maximum allowed concentration is around 0.6 mg/l.

Lead

Lead concentration should not exceed 10 µg/l. The use of lead pipes, fittings or solder can result in elevated lead levels in drinking-water, which cause adverse neurological effects. This is especially true in systems with aggressive or acidic waters. Wherever practicable, lead pipework should be replaced.

Cadmium

Cadmium concentration should not exceed 3 µg/l.

Iron

Iron levels should not exceed 200 µg/l. At levels above 300 µg/l, iron stains laundry and plumbing fixtures. There is usually no noticeable taste at iron concentrations lower than 300 µg/l, although turbidity and colour may develop.

Copper

Copper should not exceed a concentration of 2000 µg/l. Copper can stain laundry at concentrations above 1000 µg/l and, at levels above 5000 µg/l, copper can cause the water to have an orange tinge and a bitter taste. Corrosion in the piping is a typical cause of copper contamination in water.

Nickel

Nickel contamination can happen when nickel is leached from new nickel/chromium-plated taps. Low concentrations may also arise from stainless steel pipes and fittings. Nickel leaching drops off over time. Increasing the water pH to control corrosion of other materials should help to reduce nickel leaching. Nickel concentrations should not exceed 20 µg/l.

Zinc

The main corrosion problem with brasses is dezincification, which is the selective dissolution of zinc from duplex brass, leaving behind copper as a porous mass of low mechanical strength. Zinc (as zinc sulfate) imparts an undesirable astringent taste to water at a concentration of about 4000 µg/l. Water containing zinc at concentrations in excess of 3000–5000 µg/l may appear opalescent and develop a greasy film when boiled. The GDWQ do not define a health-based guideline value, but concentrations above 3000 µg/l may not be acceptable to consumers.

Hardness

Hardness, measured in concentration of calcium carbonate (CaCO3), should be between 100 mg/l (1 mmol/l) and 200 mg/l (2 mmol/l) to avoid corrosion and scaling, respectively.

Turbidity

Median turbidity should ideally be below 0.1 nephelometric turbidity units (NTU) for effective disinfection. Typical values for potable water are between 0.05 NTU and 0.5 NTU. The appearance of water with a turbidity of less than 5 NTU is usually acceptable to all consumers. High turbidity can cause material collected on the surfaces of pipes to slough off within the water distribution system.

Microorganisms

Total coliforms

Coliforms are a broad class of bacteria, and include those that can survive and grow in water. Hence, they are not useful as an index of specific faecal pathogens, but they can be used as an indicator of treatment effectiveness, and to assess the cleanliness and integrity of distribution systems, and the potential presence of biofilms. The guideline value is zero (0) coliforms per 100 ml of water.

Escherichia coli

E. coli is a type of coliform, and is considered the most suitable index of faecal contamination monitoring, including surveillance of drinking-water quality. The guideline value is zero (0) E. coli per 100 ml of water.

Intestinal enterococci

The *intestinal enterococci* group can be used as an index of faecal pollution. Most species do not multiply in water environments. Important advantages of this group are that they tend to survive longer in water environments than E. coli or thermotolerant coliforms, and are more resistant to drying and chlorination. Guideline value is zero(0) per 100 ml of water.

Clostridium perfringens

Most of these bacteria are of faecal origin and produce spores that are exceptionally resistant to unfavourable conditions in water environments, including ultraviolet irradiation, temperature and pH extremes, and disinfection processes, such as chlorination. Like E. coli, *C. perfringens* does not multiply in most water environments and is a highly specific indicator of faecal pollution. *C. perfringens*should not be present in drinking-water.

Heterotrophic plate count

Heterotrophic plate counts (HPCs) detect a wide spectrum of heterotrophic microorganisms, including bacteria and fungi. The test is based on the ability of the organisms to grow on rich growth media, without inhibitory or selective agents, over a specified incubation period and at defined temperatures (usually at 22 °C and 36 °C).

HPC is a useful parameter for the operational management of the ship's potable water system and water treatment efficacy. To properly compare HPC results, it is crucial to take more than one sample in the system. At a minimum, one sample should be taken from the tank (by use of an installed sampling tap) and another sample should be taken at the tap farthest away from the tank (usually at the bridge deck). Comparison of both (or more) samples allows interpretation of biological processes inside the distribution system and gives information about treatment efficacy. To be able to survey development of microbial growth in the particular system, it is necessary to always take the samples from the same sampling points. There are no guideline values, but the actual numbers from HPC are not as important as changes in numbers at particular locations. In distribution systems, increasing numbers from HPC can indicate a deterioration in cleanliness, possible stagnation and the potential development of biofilms.

Pseudomonas aeruginosa

P. aeruginosa is a common environmental organism and can be found in faeces, soil, water and sewage. It can multiply in water environments and also on the surface of suitable organic materials that are in contact with water. *P. aeruginosa* is a recognized cause of hospital-acquired infections with potentially serious complications. It has been isolated from a range of moist environments, such as sinks, water baths, hot-water systems, showers and spa pools. The main infection route is through susceptible tissue—notably wounds and mucous membranes—and contaminated water or surgical instruments. Therefore, at a minimum, it should be analysed in water samples taken from sick bays or hospitals, where stagnant water often poses an additional health risk.

Legionella

Legionella bacteria are the cause of legionellosis, including Legionnaires' disease. The bacteria are ubiquitous in the environment and can grow at temperatures found in piped distribution systems. The route of infection is by inhalation of droplets or aerosols; however, exposure from piped water systems is preventable through the implementation of basic water quality management measures. This includes maintaining water temperature outside the range at which *Legionella* proliferates (25–50 °C) and maintaining disinfectant residuals throughout the piped distribution system. Whenever water temperatures are found in the critical range of 25–50 °C, *Legionella* testing should be performed to assess the health risk for consumers.

Sampling procedure

Water sampling is necessary whenever technical or operational problems may exist, or when national law requires a water analysis.

The water sampling must be done by qualified personnel so that the sampling method does not influence the test results (i.e. does not contaminate the sample). Special sampling bottles and special procedures (as defined in ISO 19458) must be used.

Sample testing needs to be done using suitable methods by accredited laboratories. An internationally accepted laboratory quality standard is defined in ISO 17025. This document provides examples for parameters that are useful in certain circumstances.

An example for a reasonable microbiological sampling scheme is given below. It should be considered that the quantity of samples depends on the size of the water installation.

- Sample A: One sample should be taken from the potable water tank. This sample represents the water quality at the beginning of the ship's potable water system. Sampling should be performed as described in ISO 19458 ("purpose a"). Ship operators should be advised to install water sampling taps at the tank so that samples can be taken properly.
- Sample B: The next sample should be taken from the tap farthest from the potable water tank. It represents the influence of the distribution system. Sampling should be performed according to ISO 19458 ("purpose b").
- Sample C: If there is evidence of stagnation or other contamination in medical areas, an additional sample should be taken according ISO 19458 ("purpose c"). This sample represents the water quality for the consumer because sampling taps are not disinfected before sampling. It would be reasonable to test for P. aeruginosa at this sampling point.
- Sample D: Whenever cold water temperature is above 25 °C or hot water temperature is below 50 °C (or both), additional testing for Legionella is recommended. In this case, at least one cold and one hot water sample should be taken. It can be useful to test more sampling points (e.g. at the calorifier) to get even more information.

Document review

- Constructional drawings of potable water system.
- Drinking-water analysis reports.
- Medical logbook or gastrointestinal record book (or both).
- Water safety plan.
- Maintenance instructions of treatment devices.

References

International conventions

ILO, Accommodation of Crews (Supplementary Provisions) Convention 1970 (no. 133).

Scientific literature

Castellani PM et al. (1999). Legionnaires' disease on a cruise ship linked to the water supply system: clinical and public health implications. *Clinical Infectious Diseases*, 28:33–38.

Cayla JA et al. (2001). A small outbreak of Legionnaires' disease in a cargo ship under repair. *European Respiratory Journal*, 17:1322–1327.

Daniels NA et al. (2000). Traveler's diarrhea at sea: three outbreaks of waterborne enterotoxigenic *Escherichia coli* on cruise ships. *Journal of Infectious Diseases*, 181:1491–1495.

Goutziana G et al. (2008). *Legionella* species colonization of water distribution systems, pools and air conditioning systems in cruise ships and ferries. *BMC Public Health*, 8:390.

Joseph CA, Yadav R, Ricketts KD (2009). Travel-associated Legionnaires disease in Europe in 2007. *Eurosurveillance*, 14(18):pii:19196.

Merson MH et al. (1975). Shigellosis at sea: an outbreak aboard a passenger cruise ship. *American Journal of Epidemiology*, 101:165–175.

Mintz ED et al. (1998). An outbreak of Brainerd diarrhea among travelers to the Galapagos Islands. *Journal of Infectious Diseases*, 177:1041–1045.

O'Mahony M et al. (1986). An outbreak of gastroenteritis on a passenger cruise ship. *Journal of Hygiene (London)*, 97:229–236.

Regan CM et al. (2003). Outbreak of Legionnaires' disease on a cruise ship: lessons for international surveillance and control. *Communicable Disease and Public Health*, 6:152–156.

Rooney RM et al. (2004). A review of outbreaks of waterborne disease associated with ships: evidence for risk management. *Public Health Reports*, 119:435–442.

Guidelines and standards

Department of Health and Human Services (USA), Centers for Disease Control and Prevention, National Center for Environmental Health (2005). *Vessel sanitation program operations manual.* Atlanta, GA, and Ft Lauderdale, FL, Department of Health and Human Services (USA), Centers for Disease Control and Prevention, National Center for Environmental Health.

Department of Health and Human Services (USA), Public Health Service, Centers for Disease Control and Prevention (2005). *Vessel sanitation program construction guidelines.* Atlanta, GA, Department of Health and Human Services (USA), Public Health Service, Centers for Disease Control and Prevention.

ISO/IEC 17025:2005. *General requirements for the competence of testing and calibration laboratories.* Geneva, ISO.

ISO 19458:2006. *Water quality—sampling for microbiological analysis.* Geneva, ISO.

SO 14726:2008. *Ships and marine technology—identification colours for the content of piping systems*. Geneva, ISO.
ISO 15748-1:2002 and ISO 15748-2:2002. *Ships and marine technology—potable water supply on ships and marine structures*. Geneva, ISO.
ISO 5620-1:1992 and ISO 5620-2:1992. *Shipbuilding and marine structures—filling connection for drinking water tanks*. Geneva, ISO.
WHO (2006). *Guidelines for drinking-water quality*, first addendum to 3rd ed., Vol. 1, Recommendations. Geneva, WHO.
WHO (2007). Legionella *and the prevention of legionellosis*. Geneva, WHO.
WHO (2011). *Guide to ship sanitation*, 3rd ed. Geneva, WHO.

Code of Areas	Inspection results: evidence found, sample results, documents reviewed	Control measures and corrective actions	Required	Recommended
		9.1 Documents		
9.1.1 ❑	No water quality analysis report available, last analysis report shows contamination or not all required parameters have been analysed.	Samples have to be taken to assess actual status of potable water.Refer to WHO *Guide to ship sanitation*, Table 2.2, Examples of parameters frequently tested in potable water and typical values.	❑	
9.2.2 ❑	Medical log raises suspicion of possible waterborne diseases (e.g. diarrhoea).	Samples have to be taken to assess actual status of potable water.	❑	
9.2.3 ❑	No water safety plan available, water safety plan not adequate or no other potable water policies available to ensure safe potable water on board.	Implement a water safety plan, including all critical procedures influencing potable water quality (e.g. bunker procedure, ship water system).	❑	
		Samples have to be taken to assess actual status of potable water.	❑	
		9.2 Management		
9.2.1 ❑	Staff members not trained on safe management of the potable water system or show lack of knowledge about risks of several components.	Provide training to staff responsible for maintaining the potable water system.		❑
9.2.2 ❑	No routine checks performed to estimate the proper function of the potable water system.	Check backflow prevention devices.	❑	
		9.3 General construction on board		
9.3.1 ❑	Washbasins, showers and other taps requiring water for human consumption are connected to "fresh water" instead of potable water.	If several water systems are used, allow only fresh water to be delivered to slop sinks, laundry facilities, water closets. If non-potable water bibcock connections are used for deck flushing and cleaning purposes the tap should be labelled and secured against accidental use.	❑	
	Non-potable water is delivered to consumers in sources other than slop sinks.	Connect drinking-water used in all taps, showers, and washbasins that may supply water for human consumption to a potable water supply only.	❑	
9.3.2 ❑	Materials and pressurized components used not heat resistant.	Use only materials that are resistant to 90 °C (taps to 70 °C) so that thermal disinfection can be done.	❑	
9.3.3 ❑	Materials used not suitable for use in potable water systems. Metals and plastics not certified for use in potable water systems and may harm water quality.	Use only materials that will not contaminate the potable water with corrosive products or other substances that harm water quality. Plastics and metals in contact with potable water should be approved by the national authorities for this purpose.	❑	

Code of Areas	Inspection results: evidence found, sample results, documents reviewed	Control measures and corrective actions	Required	Recommended
9.4 Water boats and water barges				
9.4.1 ❑	Boat not properly equipped with independent potable water tanks; suitable, clean water hoses and hose fittings; or pumps and independent pipe systems to provide potable water only to the ships. Poor sanitary condition of water boat or equipment.	Install independent potable water tanks according to approved technical standards.		❑
		Equip boat or barge with suitable, clean potable water hoses that are blue (in colour) and labelled with the words "Potable water". Materials used need to meet the requirements of national health authorities.	❑	
		Clean and disinfect existing potable water hoses, fittings and equipment.	❑	
		Store all required equipment in a closed, clean, self-draining locker that is labelled with the words "Potable water hose/equipment".	❑	
		Clean and disinfect the whole storage and distribution system that is used to provide potable water to other ships.	❑	
		Remove any cross-connections to non-potable water piping or components.	❑	
9.4.2 ❑	No facilities for disinfection available, and no way to manually disinfect potable water tanks.	Equip boat or barge with suitable disinfection devices to be able to deliver chlorinated potable water to the consumer.	❑	
9.4.3 ❑	Lack of knowledge about good hygiene practices.	Train staff and develop a water safety plan to educate them about particular risks of the system.		❑
9.4.4 ❑	No actual water quality report available or report older than 3 months (this will depend on national regulations).	Order professional laboratory staff or an authorized health authority to take and analyse water samples.	❑	
9.4.5 ❑	Installation of potable water system and equipment not approved by the authorized health authority.	Call an authorized health authority to perform an audit to approve the potable water installation on board.		❑
9.5 Bunkering procedure				
9.5.1 ❑	Water from shore obviously does not meet the GDWQ	Discard contaminated water and disinfect the potable water system.	❑	
		Bunker potable water or drinking-water from a shore supply that is potable and safe.	❑	
		Verify that the potable water source meets the GDWQ before bunkering water from shore.		❑
9.5.2 ❑	No port water quality report available.	Request port water quality report before bunkering water.	❑	

Code of Areas	Inspection results: evidence found, sample results, documents reviewed	Control measures and corrective actions	Required	Recommended
9.5.3 ❑	No testing equipment available on board.	Equip ship with basic testing equipment (e.g. turbidity, pH, disinfection residuals).		❑
9.5.4 ❑	Shore supplier using inappropriate, broken or dirty materials (hoses, fittings, etc.).	Stop bunkering immediately and discard bunkered water.	❑	
		Use only own materials (e.g. hoses, fittings) that are appropriate, clean and well maintained.		❑
9.5.5 ❑	No backflow prevention to avoid contamination of the shore supply is installed in the ship, or backflow prevention is not adequately maintained or inspected.	Install backflow preventer on board to avoid any backflow from ship to the shore supply.	❑	
9.5.6 ❑	Potable water-filling connection not constructed properly to avoid connection of non-potable water hoses, not coloured blue, not tightly sealed with a cap or not secured with a corrosion-proof key lock.	Install connection flange (big ships according to ISO 5620; flange with five bolts) where only potable water hoses can be connected to avoid accidental connection to hoses carrying sewage or other non-potable liquids.	❑	
		Secure flange with a sealed cap and a corrosion-proof key lock to prevent contamination and unauthorized access.	❑	
9.5.7 ❑	Potable water filling line does not exist for every tank. Filling line is dirty, is not installed adequately, has cross-connections to other non-potable water systems, is passing through any non-potable liquid or is not labelled.	Clean and disinfect filling line.		❑
		Mark filling line with blue colour and the words "Potable water filling".		❑
		Start the filling line either in a gooseneck position downwards (ideal) or horizontally.		❑
		Place the end of the filling line at least 45 cm above the deck.		❑
		If potable water filling line is used to fill non-potable water tanks, install an air gap before installing any non-potable water tanks.	❑	
		Remove cross-connections and piping that lead through tanks carrying any non-potable liquid.	❑	
9.5.8 ❑	Filling hoses not made of suitable materials, not labelled, used for other purposes other than potable water filling, dirty, not capped, or in poor or unsanitary condition.	Equip bunker station with suitable potable water hoses, clearly marked with the words "Potable water". Normal firefighting hoses are not appropriate.	❑	
		Use potable water hoses exclusively for this purpose.	❑	
		Provide suitable hoses, approved by the authorized national administration, of at least 15 m in length and equipped with unique fittings to avoid connection to other hoses (connect to the filling line according to ISO 5620).	❑	

Code of Areas	Inspection results: evidence found, sample results, documents reviewed	Control measures and corrective actions	Required	Recommended
		Flush hoses with potable water, drain hose and raise both ends before filling with 100 mg/l of chlorine solution for 1 hour. Then drain and flush thoroughly before use or storage (with ends capped).	❑	
		Order suitable disinfection (e.g. chlorine) to disinfect hoses.	❑	
		If hoses are being blown out by compressed air, acquire a liquid trap, filter or similar device to prevent contamination via the compressed air system.	❑	
		Avoid dragging ends on the ground and dipping uncapped ends into harbour water. Train staff in good hygiene practices.	❑	
9.5.9 ❑	Potable water hose lockers do not exist, are inadequately constructed or not labelled, or are in dirty or poor condition.	Install potable water hose lockers that are made of non-corrosive, non-toxic and smooth materials; can be closed; are self draining and easily cleaned; and are labelled with words such as "Potable water hose and fitting storage".		❑
		Potable water hose locker has to be installed at least 45 cm above the deck to avoid contamination with non-potable liquids.		❑
		Label potable water hose lockers with the words "Potable water hose locker".		❑
		Maintain, clean and disinfect lockers.		❑
		Keep locker closed while not in use to avoid any contamination.	❑	
		Do not store equipment and utensils that are not required for potable water handling in potable water hose lockers.	❑	
9.5.10 ❑	During bunkering, hoses are placed directly on the ground or through harbour water.	Elevate the hoses so they are above ground. Do not touch the harbour water with the hose (to prevent cross-contamination).	❑	
		9. 6 Water production		
9.6.1 ❑	Sea chest for freshwater intake located on the same side, and at the same section or aft of the sanitary overboard discharge.	Place the sanitary overboard discharge and the fresh water intake on opposite sides.		❑
		If it is on the same side, place the sanitary discharge as far above the freshwater intake as possible.		❑
		Never produce potable water while sanitary overboard water is being discharged.	❑	

Code of Areas	Inspection results: evidence found, sample results, documents reviewed	Control measures and corrective actions	Required	Recommended
9.6.2 ❑	Water has been produced in unsafe areas such as in ports, on rivers or on anchorage.	Take samples from the tank and, at the minimum, perform a microbiological analysis (e.g. test for E. coli, coliforms, enterococci, C. perfringens, HPC) to evaluate level of risk.	❑	
		Perform disinfection of piping and components between evaporator and tank.		❑
		Inform responsible staff that water production in unsafe areas is forbidden.	❑	
9.6.3 ❑	Evaporator process temperature below 80 °C and no disinfection device installed to treat the distillate.	Control temperature.	❑	
		Install disinfection device (e.g. automatic chlorination) behind evaporator and control temperature.	❑	
9.6.4 ❑	Evaporator has no opening and cannot be maintained or inspected.	Equip ship only with components that can be maintained and inspected.		❑
9.6.5 ❑	Information about evaporator missing; no manufacturer contact details or maintenance instructions available.	Obtain and post the most important instructions, including technical specifications, close to the evaporator.	❑	
9.6.6 ❑	Salinity sensor not installed or working. "Automatic discharge to waste if distillate salty" not installed or working.	Repair or install a low-range salinity sensor with alarm function, and an automatic switch-off or discharge.		❑
9.6.7 ❑	No sampling tap at the distillate outlet.	Install a heat-resistant metal sampling tap so distillate sampling can be done.		❑
9.6.8 ❑	Reverse osmosis: staff members show a lack of knowledge about health risks of membrane breaches and device maintenance.	Conduct training for staff who are responsible for the potable water system.		❑
		9. 7 Treatment components		
9.7.1 ❑	Piping installed that allows bypassing of treatment components (e.g. disinfection devices, filters, rehardening systems).	Remove all bypasses.	❑	
9.7.2 ❑	Remineralization process not working or not existing behind evaporators or reverse-osmosis devices. Insufficiently treated distillate or permeate delivered to the consumer.	Install remineralization devices to control corrosivity of the water.		❑
		Inform consumer about the water quality (e.g. hardness, pH).		❑
9.7.3 ❑	Remineralization devices not cleaned, maintained or refilled regularly. Maintenance procedures unknown	Train staff on maintenance of remineralization devices.		❑
		Empty, clean, disinfect and refill the remineralization device according to manufacturer's instructions.		❑
		Label device with information about manufacturer and technical specifications.	❑	

Code of Areas	Inspection results: evidence found, sample results, documents reviewed	Control measures and corrective actions	Required	Recommended
9.7.4 ❑	Filters dirty or maintenance procedures unknown.	Backwash or replace filter medium according to the manufacturer's instructions.		❑
9. 8 Disinfection				
9.8.1 ❑	No disinfection device installed to treat produced water downstream of the desalination device.	Install automatic disinfection device (preferably chlorination) behind the seawater desalination devices (evaporator or reverse-osmosis device) according to approved technical standards. The water must be at least 30 minutes in the tank to disinfect properly.	❑	
9.8.2 ❑	Disinfection of bunkered water not possible. No technical ability to add disinfection measures during bunkering procedure.	Install automatic disinfection devices (preferably chlorination in the bunker line) to treat bunkered water according to approved technical standards.	❑	
9.8.3 ❑	Manual chlorination performed without sufficient knowledge.	Install an automatic chlorination device.		❑
		Order a professional company to perform chlorination of the water. Respect the guideline values of national health authorities.	❑	
9.8.4 ❑	Continuous halogenation performed, but no continuous recording of the halogen concentration installed.	Record disinfection residue at the farthest point in the system where significant water flow exists (e.g. bridge deck).	❑	
		Check chlorine concentration (free and total) and pH of the water before and after each bunkering, and at regular intervals during normal operation.		❑
9.8.5 ❑	There is no chlorine in the system during chlorination.	Free chlorine level should be between 0.5 mg/l and 1.0 mg/l during disinfection (consider national standards) and at the point of delivery, the minimum should be 0,2 mg/l.		❑
9.8.6 ❑	A sampling cock is missing downstream of the disinfection unit.	Install sampling cock 3 metres downstream of the disinfection injection device.		❑
9.8.7 ❑	Quantity of stored disinfection chemicals insufficient to treat the water volume over the whole duration of travel.	Store enough disinfection chemicals to be able to perform continuous water disinfection.	❑	
9.8.8 ❑	Chlorine and pH testing kits not available.	Equip ship with a testing kit for free and total chlorine (range 0–5 mg/l) and pH (range 6–10).		❑
9.8.9 ❑	Potable water system needs a superchlorination, due to contamination, or system repair or maintenance.	Perform professional superchlorination with 50 mg/l over 24 hours. Water is unfit for consumption in the meantime.	❑	
9.8.10 ❑	UV system not well maintained. Lack of knowledge concerning operation and maintenance. Bypass piping installed. System not approved by the national authority. No spare parts available on board.	Clean and disinfect the UV lamp according to the manufacturer's instructions.	❑	
		Arrange to have the necessary spare parts (e.g. UV lamp) on board.		❑
		Remove any bypass construction around disinfection devices.	❑	

Code of Areas	Inspection results: evidence found, sample results, documents reviewed	Control measures and corrective actions	Required	Recommended
		During the next inspection, provide a written approval for the system, or install a new device that is approved by the national authority.	❑	
		Label device with manufacturer's information and technical specifications.	❑	
9. 9 Tanks				
9.9.1 ❑	Potable water tanks not identifiable or volume not labelled.	Clearly identify potable water tanks (e.g. with the words "Potable water" in large print).	❑	
		Label tanks with the volume.	❑	
9.9.2 ❑	Tank location does not meet the following requirements: • The tank is located in areas without exposure to heat, dirt, vectors or other contamination. • The tank is protected from any contamination from outside the tank. • The tank is independent of the ship hull or the bottom of the tank is at least 60 cm above the maximum load water line. • The cofferdam is at least 45 cm above and between tanks that are not for potable water storage, and between potable water tanks and the hull. If a deck is the top of a potable water tank, no access or opening to the tank is allowed via the deck.	Potable water tanks should be located in areas where they will be at no risk for contamination and are protected against contamination from other sources.	❑	
		Protect the water from heating (to between 25 °C and 50 °C) to prevent microorganism growth.	❑	
9.9.3 ❑	Drainage lines or piping carrying non-potable liquids (e.g. sewage or fuel) are leading through the potable water tank.	Remove piping that is passing through potable water tanks or construct an acceptable tunnel.	❑	
9.9.4 ❑	Lines carrying sewage or other highly contaminated liquids are passing directly over maintenance openings of potable water tanks.	Remove maintenance openings or piping to avoid accidental cross-contamination by leaking pipes.	❑	
9.9.5 ❑	Cross-connections: there are inadequate lines to divert potable water by valves or interchangeable pipe fittings to other systems (e.g. connection to firefighting system).	Remove any cross-connections between potable and non-potable water tanks and pipes.	❑	
		Where removal of cross-connection is not possible, approved backflow-prevention devices need to be installed.		❑

Code of Areas	Inspection results: evidence found, sample results, documents reviewed	Control measures and corrective actions	Required	Recommended
9.9.6 ❑	Sewage system components, soil-waste drains or solid-waste facilities are installed directly above potable water tanks or tank maintenance openings so that there is a risk of contamination by dirt or spillage. Toilets and bathroom spaces extend over any part of a deck that forms the top of a tank used for potable water or wash water.	Remove toilet or solid-waste facilities that are located directly above potable water tanks or maintenance openings.		❑
9.9.7 ❑	Tank capacity does not ensure an independent water supply of at least 2 days without bunkering or production of new water.	Install potable water tanks of sufficient size.		❑
9.9.8 ❑	Potable water tanks share a common wall with the hull or other non-potable holding tanks.	Construct potable water tanks so that they do not share a common wall with the hull or other non-potable holding tanks. A cofferdam of at least 45 cm should be between them.		❑
		Install a conductivity sensor and monitoring system that triggers an alarm and automatic closing of valves in case of contamination from salt water or other liquids with high conductivity.	❑	
9.9.9 ❑	Tanks do not have a maintenance opening for inspection.	Install a maintenance opening that gives access for cleaning, repair and inspection, preferably located on the side of the tank(s).	❑	
		Maintenance openings at the top of a tank need to have a coaming or curb that is raised at least 1.25 cm above the top of the tank.		❑
		All access to potable water tanks need to be kept tightly closed when not being used.	❑	
9.9.10 ❑	Potable water tanks not constructed or coated for potable water contact. Tank coating is rendering water unfit for human consumption.	Store potable water only in tanks constructed for this purpose, and protect against any contamination from outside or within the tank (e.g. corrosion, or inappropriate or faulty tank coating).		❑
		Use potable water tanks that are constructed of metal or another suitable material safe for contact with water.	❑	
		Use potable water tanks and pipes that are constructed and coated with safe and durable material.		❑
		Supply written documentation that tank coating is approved for potable water tanks and that all manufacturers' recommendations have been followed, or take samples for chemical analysis (or both).		❑

Code of Areas	Inspection results: evidence found, sample results, documents reviewed	Control measures and corrective actions	Required	Recommended
9.9.11 ❑	Tank ventilation or overflow (or both) does not exist, is connected to non-potable water tanks, or does not prevent entry of contaminants or vectors.	Terminate the vent or overflow with the open end pointing down, either inside the vessel (above bilge level) or at least 45 cm above a weather deck in a sheltered place.	❑	
		Screen vents and overflow pipes with corrosion-resistant mesh of mesh count 16 × 16 or finer.	❑	
		Use vents or overflow that is at least the same diameter as the filling line.		❑
		Remove any direct connections between ducts from potable water tanks and non-potable liquid tanks.	❑	
9.9.12 ❑	Tank not completely drainable, or drain insufficient in diameter, location or construction.	Install a sufficient drain to allow complete drainage of the tank.		❑
		Ensure the distance between the end-of-drain line and the highest point of the bilge is more than 45 cm.	❑	
9.9.13 ❑	Tank level indicators posing a risk of contamination.	Do not use dip sticks or sounding lines for sounding of tank levels.	❑	
		Equip tanks with level indicators like water-gauge glass, petcocks, enclosed float gauges or water-operated pressure gauges.		❑
9.9.14 ❑	No sample cocks installed at the tanks.	Install heat-resistant sample cocks at each tank to allow water-quality testing. The sample cocks need to point downwards, and be identified by numbers.	❑	
9.9.15 ❑	No regular cleaning of the tanks has been performed, or sediments have been found at the bottom of the tank.	Inspect, clean, flush and disinfect the tank every 6 months. Document the measure in the logbook.	❑	
9.9.16 ❑	After entering tanks for repairs or maintenance, no cleaning measures have been performed.	Perform post-repair disinfection.	❑	
		Enter tanks wearing only clean, single-use overalls, clean rubber boots (used for this purpose only), a face mask and rubber gloves to reduce risk of contamination.		❑
		9.10 Potable water pumps		
9.10.1 ❑	Potable water pumps used for liquids other than potable water.	Do not use potable water pumps for any liquid other than potable water.	❑	
		Install potable water pumps that are approved for this purpose.	❑	
	Pumps not able to establish a permanent positive pressure in the system.	Install a potable water pressure tank if no permanent pressure is available in the distribution system.	❑	
	No standby pump available for emergencies.	Equip the system with an emergency pump that always allows access to drinking-water.		❑

Code of Areas	Inspection results: evidence found, sample results, documents reviewed	Control measures and corrective actions	Required	Recommended
9.10.2 ❑	Hand pumps not installed in a way that prevents contamination of the potable water.	Install hand pumps in a way that prevents contamination.		❑
9.11 Potable water-pressure tank				
9.11.1 ❑	Potable water-pressure tank (hydrophor tank) cross-connected to non-potable water systems with a compressed air line without using an appropriate fail-safe device.	Install an independent compressor.		❑
		Install a press-on valve with a liquid trap of suitable size in the supply line.	❑	
9.11.2 ❑	Manufacturer's maintenance instructions missing or not respected.	Label tank with specifications and manufacturer's information.	❑	
		Perform regular maintenance according to manufacturer's instructions (e.g. cleaning and disinfection).	❑	
9.11.3 ❑	No sampling cock installed at the water-pressure tank.	Install a suitable sampling cock at the water-pressure tank to take water samples for analysis.		❑
9.12 Calorifier and hot-water system				
9.12.1 ❑	Size of the calorifier is inappropriate (based on the calculated hot-water consumption).	Install larger calorifier to support all people on board sufficiently.	❑	
		Install smaller calorifier or install a decentralized hot-water supply to avoid unnecessary complications (suitable for smaller boats).	❑	
9.12.2 ❑	Calorifier or hot-water piping's material is not suitable and is making water unfit for human consumption (e.g. due to corrosion or leaching of chemicals). Calorifier not insulated.	Install components that do not harm the water quality.	❑	
		Install thermal insulation around the calorifier.	❑	
9.12.3 ❑	No maintenance opening, or a lack of maintenance knowledge.	Install maintenance opening, if possible.		❑
		Renew calorifier.		❑
		Clean, decalcificate and disinfect the calorifier.		❑
9.12.4 ❑	Temperatures in an inappropriate range or thermometers missing.	Set temperature in the calorifier outlet to 65 °C.	❑	
		Set temperature in the return line (circulation systems only) above 55 °C.	❑	
		Equip calorifiers with thermometers to check temperatures in the outflow, boiler and return lines.	❑	

Code of Areas	Inspection results: evidence found, sample results, documents reviewed	Control measures and corrective actions	Required	Recommended
9.12.5 ❑	Hot-water circulation pumps not permanently running or not approved for drinking-water systems.	Run the hot-water circulation pump permanently to avoid stagnation and cooling down of the water in the piping.	❑	
		Use only pumps that are approved for drinking-water.	❑	
9.12.6 ❑	Hot and cold water piping laid side-by-side without thermal insulation.	Insulate cold-water and hot-water piping to avoid growth of microorganisms (e.g. Legionella).		❑
		9.13 Plumbing		
9.13.1 ❑	Piping made of inappropriate material (e.g. lead- or cadmium-lined pipes).	Replace all pipes, fittings and joints that are lead or cadmium lined, or otherwise not appropriate for potable water contact, with adequate piping.	❑	
9.13.2 ❑	Piping not clearly identified as potable water piping (e.g. it does not have blue stripes every 5 m along the pipe).	Paint all potable water piping with colour code according to ISO 14726.	❑	
9.13.3 ❑	Piping contains unnecessary dead legs where water stagnates.	Remove dead legs, and unnecessary taps and faucets.		❑
		Regularly flush dead legs according to a flushing schedule.	❑	
9.13.4 ❑	Potable water piping is passing through or connected to sewage tanks, sewage piping or other tanks containing non-potable liquids.	Remove direct connections between potable water and non-potable water systems. Install air gaps or other appropriate backflow prevention to avoid cross-contamination.	❑	
	Potable water piping is leading through the bilge.	Rearrange piping wherever potable water lines are leading through non-potable liquids (e.g. bilge).		❑
9.13.5 ❑	Backflow preventers missing where potable water system is connected to non-potable water systems under pressure. Most important connections to check are: • supply lines to swimming pools, whirlpools, hot tubs, bathtubs, showers and similar facilities; • photography laboratory developing machines; • beauty- and barber-shop rinse hoses; • garbage grinders;	Install appropriate backflow prevention. Choose type of backflow prevention (e.g. air gap, vacuum breaker) according to the particular risk.	❑	

Code of Areas	Inspection results: evidence found, sample results, documents reviewed	Control measures and corrective actions	Required	Recommended
9.13.5 ❑	• hospital and laundry equipment; • air-conditioning expansion tanks; • boiler-feed water tanks; • fire systems; • toilets and bidets; • freshwater or saltwater ballast systems; • bilge or other wastewater locations; • international shore connection.		❑	
9.13.6 ❑	Backflow preventers not adequately maintained; no inspections or testings have been performed or documented.	Perform testing of installed backflow-prevention devices at least once per year. Record the results so that the testing logs can be shown to the ships inspector.	❑	
9.13.7 ❑	Materials used in the piping systems not heat resistant up to 90 °C.	Use only materials that are heat resistant up to 90 °C so that thermal disinfection can be done.		❑
		9.14 Taps, faucets and showerheads		
9.14.1 ❑	Faucet filters or other types of terminal filters are used to improve water quality, but not replaced or maintained regularly.	Do not use terminal filters regularly without maintenance and renewal.		❑
		Ensure that used filter devices meet the criteria of the national health administration or the local health authority.		❑
9.14.2 ❑	Water outlets not labelle	Label potable water outlets with the words "Potable water".	❑	
		Label non-potable water outlets with the words "Unfit for human consumption or use!"	❑	
9.14.3 ❑	Fixtures and fittings corroded or dirty (or both).	Ensure that fixtures and fittings are made of a material resistant to the corrosive effects of salt water and saline atmosphere; they should have rounded internal corners to facilitate cleaning.	❑	
9.14.4 ❑	Showerheads or aerators (or both) dirty or otherwise in poor condition.	Clean and disinfect aerators and showerheads.	❑	
		Replace aerators and showerheads that are in poor condition.		❑
9.14.5 ❑	Hot-water temperature at any tap below 50 °C.	Increase hot-water temperature in the calorifier to avoid growth of Legionella.	❑	
		Perform a hot-water flush (i.e. all taps and showerheads turned on consecutively at a temperature of 70 °C for more than 3 minutes).		❑
		Sample the water to assess the risk of Legionella contamination.	❑	

Code of Areas	Inspection results: evidence found, sample results, documents reviewed	Control measures and corrective actions	Required	Recommended
9.14.6 ❑	Cold-water temperature at any tap above 25 °C.	Insulate the cold-water system, and avoid any exposure to excessive heat or other potable water components.		❑
		Sample the water to assess the risk of *Legionella* contamination.	❑	
9.15 Hand-washing facilities				
9.15.1 ❑	Washbasin or other places where water for human consumption is offered does not have potable water.	Deliver potable water to all washbasins, bathtubs, showers and other places where water is used for human consumption.	❑	
9.16 Drinking-water fountains				
9.16.1 ❑	Water-jet orifices not slanted or orifice not protected by cover. Insufficient flow to ensure the mouth does not come in contact with the fountain. Flow of stream cannot be controlled by user. Evidence of mould or slime build-up.	Protect orifice with a cover.		❑
		Increase flow to avoid mouth contact with the fountain.		❑
		Clean and disinfect the whole fountain (including inner parts).	❑	
		Disconnect the drinking-water fountain from the potable water system while cleaning and disinfecting, then reconnect.		❑
9.17 Water-service containers				
9.17.1 ❑	Coolers that permit direct contact between ice, water and coolers (in which water bottles are inserted neck downwards into the cooling chamber) are used.	Do not use these devices.		❑
9.18 Ice machines				
9.18.1 ❑	Non-potable water used to make ice cubes or other ice for drinking purposes. Ice machine or ice-bin interior dirty or poorly maintained.	Do not make ice for human consumption from non-potable water.	❑	
		Clean and disinfect ice machine.	❑	
9. 19 Criteria for taking water samples (only some possible reasons)				
9.19.1 ❑	Water has been produced in unsafe water bodies (e.g. rivers, lakes, anchorage).	Test for microorganism contamination (especially E. coli, coliforms, enterocci, HPC, *C. perfringens*).	❑	
9.19.2 ❑	Water has been bunkered from unsafe sources (e.g. using dirty hoses).	Test for microorganism contamination (especially *E. coli*, coliforms, enterocci, HPC, *C. perfringens*, *P. aeruginosa*).	❑	

Code of Areas	Inspection results: evidence found, sample results, documents reviewed	Control measures and corrective actions	Required	Recommended
9.19.3 ❑	No halogen or chlorine residue measured when testing.	Test for microorganism contamination (especially E. coli, coliforms, enterocci, HPC).	❑	
9.19.4 ❑	Potable water installation does not meet the international or national technical standards.	Sample for microorganism and chemical contamination based on risk stratification and national health regulations.	❑	
9.19.5 ❑	Water temperatures out of range (cold water >25 °C or hot water <50 °C).	Test for microorganism contamination (especially *Legionella* spp.).	❑	
9.19.6 ❑	Stagnant water found, or poorly maintained aerators and showerheads (especially in medical areas).	Test for microorganism contamination (especially *P. aeruginosa*, HPC).	❑	
9.19.7 ❑	Chemical odour or taste.	Test for chemical contamination (e.g. tank coating, fuel).	❑	
9.19.8 ❑	Coloured water.	Test for chemical contamination (e.g. pipe corrosion).	❑	

Area 10 Sewage

Introduction

Large amounts of wastewater can accumulate on ships, depending on the number of people on board, type of ship and duration of voyage. This wastewater can be separated into grey water (wash water, shower, etc.) and black water. "Sewage" and "black water" are often used interchangeably.

According to the internationally accepted definition in the IMO International Convention for the Prevention of Pollution from Ships 1973 (modified by the protocol of 1978 relating thereto [MARPOL 73/78]), sewage is defined as:

- drainage and other wastes from any form of toilets, urinals and WC flushing system;
- drainage from medical premises (e.g. dispensary, sickbay) via washbasins, wash tubs and scuppers located in such premises;
- drainage from spaces containing living animals (e.g. livestock carriers); or
- other wastewaters (e.g. grey water from showers) when mixed with the drainages defined above.

Sewage is one type of wastewater, and is a major actual or potential source of potable water pollution with infectious agents. Pollution can also come from many chemical characteristics, including high concentrations of ammonium, nitrate and phosphorus; high conductivity (due to high dissolved solids); and high alkalinity, with pH typically ranging between 7 and 8. Trihalomethanes are also likely to be present as a result of past disinfection.

Main risks

Unsafe management and disposal of sewage can readily lead to adverse health consequences. Black water may harbour many different harmful substances, such as chemicals, pharmaceuticals and biological agents. It is well known that pathogenic amoebae, bacteria, viruses, worms, fungi and parasites survive in untreated black water. The main risk is disease spread by contaminated and insufficiently treated sewage that has been discharged into the surrounding water. Cross-contamination of potable water, accidents (e.g. leakage or overflow) and acquired infections during maintenance work are some of the additional health risks.

The most common method of treating sewage is to flush sewage from toilets through a piping system into a holding tank where it is comminuted, decanted and broken down by naturally occurring bacteria in an aerobic process. It is then disinfected before it is discharged into the open sea. It is important to consider that an excessive use of cleaners and disinfectant in the sewage system may destroy the natural bacteria in the treatment plant. The aerobic process needs oxygen; therefore, aerators blow air into the biological compartment. Toxic gases can be produced during this process.

International standards and recommendations

The IMO's International Convention for the Prevention of Pollution from Ships 1973 (modified by the protocol of 1978 relating thereto [MARPOL 73/78]), Annex IV: Prevention of pollution by sewage from ships, entered into force on 27 September 2003. It was revised on 1 April 2004 and the revision entered into force on 1 August 2005.

Since September 2008, MARPOL 73/78 Annex IV and the Marine Environment Protection Committee's (MEPC's) amendment MEPC.115(51) state that all ships engaged in international voyages and of a size >400 gross tonnage, or certified to carry >15 persons, need to be equipped with at least one of the following sewage systems:

- sewage holding tank with sufficient capacity and visual level indicator;
- sewage comminuting and disinfecting system, including storage tank;
- sewage treatment plant that is approved according to MEPC.2(VI), Recommendation on international effluent standards and guidelines for performance tests for sewage treatment plants.

Passenger ships deal with large amounts of wastewater and are often operating in protected areas. Therefore, onboard cruise ship treatment plants often use the principles of membrane filtration or reverse osmosis (or both).

In MARPOL 73/78 Annex IV, different regulations concerning handling of sewage are defined. IMO's MEPC has defined additional amendments with more detailed information:

- MEPC.2(VI), Actual criteria for testing of sewage treatment plants;
- MEPC.115(51), Revision of the regulations in MARPOL Annex IV;
- MEPC.157(55), Standards for the rate of discharge of untreated sewage;
- MEPC.159(55), Criteria for sewage treatment plants built after January 2010

Document review

- Technical drawings of sewage system.
- IMO International Sewage Pollution Prevention (ISPP) certificate.
- International Safety Management (ISM) manual.
- Sewage management plan (if available).
- Maintenance instructions of sewage treatment plant (if installed).

References

International conventions

ILO, Maritime Labour Convention 2006.

IMO, International Convention for the Prevention of Pollution from Ships 1973. (modified by the protocol of 1978 relating thereto [MARPOL 73/78]).

Scientific literature

Rooney RM et al. (2004). A review of outbreaks of foodborne disease associated with passenger ships: evidence for risk management. *Public Health Reports*, 119:427–434.

Guidelines and standards

ISO 14726-2:2008. *Ships and marine technology—Identification colours for the content of piping systems—Part 2: Additional colours for different media and/or functions.* Geneva, ISO, 2009.

WHO (2011). *Guide to ship sanitation*, 3rd ed. Geneva, WHO.

Code of Areas	Inspection results: evidence found, sample results, documents reviewed	Control measures and corrective actions	Required	Recommended
10.1 Document review				
10.1.1 ❑	ISPP certificate is more than 5 years old, does not exist, or is issued in neither English nor French.	Conduct renewal survey of the whole sewage system and obtain a new ISPP certificate. Inform the competent authority (e.g. Port State Control).	❑	
10.1.2 ❑	Regulations for familiarization of the sewage treatment plant operation for ship personnel do not exist.	Incorporate instructions for operation and maintenance of the sewage system into ISM manual.		❑
10.1.3 ❑	No technical drawings available.	Prepare technical drawings for the next inspection.		❑
10.1.4 ❑	No operation and maintenance instruction manual available.	Provide maintenance instructions for the next inspection.	❑	
10.1.5 ❑	Sewage not included in the waste management plan, or no waste management plan exists.	Develop a waste management plan that includes procedures for the handling of sewage.		❑
10.2 Sewage from the galley or pantries				
10.2.1 ❑	Sink drain (for food preparation or dishwashing) directly connected to the wastewater system.	Install sufficient backflow prevention (e.g. air gaps).		❑
		Install signs that advise strict cleaning of the sink before any food is prepared there.		❑
10.2.2 ❑	Drain pipes of ice machines, dishwashing machines or food refuse grinders directly connected to wastewater system, and not equipped with appropriate backflow prevention.	Install sufficient backflow prevention (e.g. air gaps) at dishwashing machine.		❑
		Install sufficient backflow prevention (e.g. air gaps) at ice machines.		❑
		Install air gaps or mechanical backflow preventers at garbage grinders and food-waste systems.		❑
10.2.3 ❑	No grease interceptors installed between galley wastewater system and sewage system; or grease interceptors are overflowing, dirty or otherwise insufficiently maintained.	Install grease interceptors between the galley wastewater system and the sewage system.		❑
		Clean grease interceptors regularly and dispose of collected grease.		❑

Code of Areas	Inspection results: evidence found, sample results, documents reviewed	Control measures and corrective actions	Required	Recommended
10.3 Sewage from medical areas				
10.3.1 ❑	Technical drawing or physical inspection of the sewage system shows that liquids from medical areas (e.g. drains, showers, bathtubs, washing bowls, toilets) are not drained into the sewage system.	Change the construction so that all sewage from medical areas goes into the sewage system.	❑	
10.4 Sewage from public and common bathrooms, hand-wash stations and toilets				
10.4.1 ❑	Insufficient deck drains connected to the sewage system.	Install sufficient deck drainage that is connected to the sewage system.		❑
10.4.2 ❑	Sewage and bathwater are clogged, or there is visible or reported backflow.	Clean and unclog the drains and pipes.		❑
		Install piping of adequate size.		❑
10.4.3 ❑	Toilet or flushing system not working properly.	Repair toilet or flushing system (or both).	❑	
10.4.4 ❑	Technical drawing or physical inspection of the sewage system shows that bathrooms or toilets are located on a deck that forms the top of potable water tanks.	Close affected bathroom(s) or toilet(s). Disconnect all piping (sewage and potable water) close to main pipes to avoid harmful stagnation.		❑
10.5 Cargo holds				
10.5.1 ❑	Cargo-hold drainage or refrigerated cargo spaces are directly connected to the sewage system.	Connect these drains to a common drainage system that is separate from any sewage system.		❑
10.6 Animal excrement				
10.6.1 ❑	No drains of sufficient size installed and connected to a holding tank or a treatment system.	Install suitable drainage to avoid any pooling or spilling of animal excrement.		❑
		Install a holding tank with the appropriate capacity to store sewage until ship can safely discharge the sewage.		❑
10.7 Piping system				
10.7.1 ❑	Colour coding missing at the piping (e.g. in black–blue–black every 5 m).	Suitable colour coding has to be painted onto the piping at least every 5 m (e.g. black–blue–black according to ISO 14726:2008).	❑	
10.7.2 ❑	Pipes not well maintained; they are clogged, of inadequate size or leaking	Clean and maintain the piping.		❑
		Repair leakage immediately.	❑	
		Install piping of adequate size.		❑
10.7.3 ❑	Cross-connections to other systems containing liquid are found, or sewage pipes are leading through potable water tanks or are connected to potable water piping.	Remove all cross-connections.	❑	
		Remove all piping in the tanks that do not contain potable water.	❑	

Code of Areas	Inspection results: evidence found, sample results, documents reviewed	Control measures and corrective actions	Required	Recommended
10.7.4 ❑	Drain lines carrying sewage and grey water are passing directly over: • galleys, buffets or bars; • food preparation or serving areas; • food equipment or utensil washing areas; • food storage areas.	Change piping arrangement; drain lines should not pass over critical areas.		❑
	10.8 Sewage holding tank (mandatory for all ships >400 tons gross tonnage or with ≥15 persons on board)			
10.8.1 ❑	No sewage holding tank available.	Install sewage holding tank in isolated position (cofferdam and coaming), and of sufficient size and material. Equip with a level indicator, high level alert, cleaning access and an overflow system.	❑	
10.8.2 ❑	*Identification and capacity* Tank is not identified, or the tank capacity: • is unlabelled; • is insufficient; • does not correspond to the ISPP certificate	Place a sign on the tank and label with the words "Sewage-holding tank".	❑	
		Label tank clearly with the correct tank capacity in m^3.	❑	
		Install holding tank with sufficient capacity (at least 114 liter per person per day, or according to the Helsinki Commission [HELCOM] Recommendation 10/11).	❑	
		Renew ISPP certificate. Inform the authorized authority (e.g. Port State Control).	❑	
10.8.3 ❑	*Leakage and overflow* Tank not in an isolated position (no cofferdam) or sharing a common wall with potable water tanks. Tank not secured against overflow or leakage by coaming. Tank has no visual level indicator, or no high-level alert installed (or both). Evidence found for leakage or overflow (e.g. the holding tank or surrounding area is dirty).	Install tank in an isolated position that does not share a common wall with potable water tanks. Surround tank with a cofferdam.		❑
		Install coaming around tank to avoid spread of spillage.	❑	
		Install easily visible level indicator on the outside of the tank.	❑	
		Install a high-level alert.		❑
		Clean and disinfect the contaminated area.	❑	
		Install tank with sufficient capacity (if the reason for overflow is insufficient tank capacity).		❑
		Repair leaks.	❑	
10.8.4 ❑	No or inadequate ventilation of the tanks.	Install a vent on the tank.		❑
		Lead emissions to the outside of the ship, away from any air intakes.	❑	

Code of Areas	Inspection results: evidence found, sample results, documents reviewed	Control measures and corrective actions	Required	Recommended
10.9 Comminuting and disinfection plant (if installed)				
10.9.1 ❑	Comminuting and disinfection system has no type approval, is more than 5 years old, or is not displaying the test results of the sample taken after disinfection, or the ISPP certificate is not available.	Install type-approved system.	❑	
		Renew ISPP certificate. Inform authorized authority.	❑	
		Take samples to determine the efficiency of the disinfection unit.	❑	
10.9.2 ❑	*Identification and capacity:* Plant not identifiable. Tank capacity: • is not labelled; • is insufficient to hold comminuted and disinfected sewage while the ship is less than 3 nautical miles away from nearest land; or • does not correspond to the ISPP certificate.	Install an additional holding tank to store disinfected and comminuted effluent of the treatment plant.	❑	
		Make tank clearly identifiable with sign, and label with the words "Comminuting and disinfection plant".	❑	
		Label tank clearly with the correct tank capacity in m^3.	❑	
10.9.3 ❑	Not enough disinfectant to operate the system during the next voyage.	Store enough disinfectant to operate the treatment system for at least two complete voyages.		❑
10.9.4 ❑	Ship personnel have insufficient knowledge about operation and maintenance of the comminuting and disinfection plant.	Train technical staff in the operation and maintenance of the sewage system.		❑
		Train staff in the operation and maintenance of the sewage system according to the ISM manual.		❑
10.10 Sewage treatment plant (if installed)				
10.10.1 ❑	Sewage treatment plant has no type approval, is more than 5 years old or is not displaying test results, or the ISPP certificate is not available.	Install type-approved system.	❑	
		Inform authorized authority to renew the ISPP certificate.	❑	
10.10.2 ❑	Treatment plant in inoperable condition or bypassed.	Turn on sewage treatment plant.	❑	
		Store all sewage in a suitable holding tank, or discharge sewage to an official port reception facility.	❑	
		Repair or maintain sewage treatment plant before discharging treated sewage into waterbodies or the open sea.	❑	

Code of Areas	Inspection results: evidence found, sample results, documents reviewed	Control measures and corrective actions	Required	Recommended
10.10.3 ❑	Treatment plant appears to be in an unsatisfactory condition; for example: • it is switched off or bypassed; • the aerators for the biological compartment are not working; • there is too much sludge in the compartments; • there is no disinfectant available; • the treated effluent shows visible pieces; • there is a strong, conspicuous smell or explicit colour (note: a slightly brown or yellow colour is harmless); • the drain valve for the biofilter or settling tank is open or broken.	Turn on sewage treatment plant; treat all sewage before discharge.	❑	
		Perform maintenance works according to manufacturer's instructions to obtain required effluent standard.	❑	
		Do not discharge effluent closer than 12 nautical miles to the coastline until all problems are fixed.	❑	
		Repair the aerators.	❑	
		Refill with suitable disinfectant according to manufacturer's instructions.	❑	
		Take samples for microorganism analysis according to MEPC regulation.	❑	
		Remove excess sludge from the tanks and discharge it to an official port reception facility.	❑	
		Clean, inspect and repair the tanks according to the manufacturer's instructions, when required (e.g. every 6 months).		❑
10.10.4 ❑	No sampling point available or it is not suitable for sampling.	Install metal sampling point at treated effluent outlet.	❑	
10.10.5 ❑	No or inadequate ventilation of the tanks.	Install a vent on the tank.		❑
		Direct emissions to the outside of the ship, away from any air intakes.	❑	
10.11 Discharge				
10.11.1 ❑	Untreated sewage has been discharged into the port basin, the river or another protected area.	Stop discharging immediately and report the incident to the port authority	❑	
10.11.2 ❑	Overboard valve open or broken.	Close or repair valve immediately.	❑	
10.11.3 ❑	Pipes that may discharge untreated sewage, residues from sewage treatment and sewage from holding tanks directly into the port basin are not closed.	Close all pipe valves immediately.	❑	
10.11.4 ❑	Excess sludge from tanks or treatment plants is not appropriately stored for eventual disposal to land-based facilities or open sea.	Store excess sludge in appropriate tanks until eventual discharge to port-reception facilities or open sea.		❑

Code of Areas	Inspection results: evidence found, sample results, documents reviewed	Control measures and corrective actions	Required	Recommended
10.11.5 ❑	MARPOL 73/78 discharge rules not known, or national discharge rules for the actual port or surrounding waters not known.	Train appropriate personnel in MARPOL 73/78 regulations or national port regulations, and implement these procedures into the ISM manual.		❑
10.11.6 ❑	Pipeline suitable for the discharge of sewage to a reception facility is missing; the discharge shore connection does not meet requirements of MARPOL 73/78 Annex IV; or a quick-closing coupling is used but has not been agreed to by the administration.	Install suitable pipeline to allow sanitary discharge of sewage to port reception facilities.	❑	
		Install discharge connections with standard flange(s) according to MARPOL 73/78 Annex IV.	❑	
10.12 Discharge hoses				
10.12.1 ❑	Sewage discharge hoses not available; hoses not made of durable, impervious material with a smooth interior surface; or not labelled as sewage discharge hoses.	Order suitable sewage discharge hoses, made of durable, impervious material, with a smooth interior surface.		❑
		Label sewage discharge hoses with the words "For waste discharge only" to avoid accidental cross-contamination.		❑
10.12.2 ❑	Hoses are in dirty condition; they appear not to be cleaned and disinfected. Hoses are not stored in a designated convenient place, or are stored together with potable water equipment.	Clean, flush and disinfect sewage discharge hoses after each use.		❑
		Store sewage discharge hoses at a dedicated place, labelled with the words "Waste discharge hose".		❑
		Clean and disinfect the storage place.		❑
		Do not store any potable water hose or equipment together with sewage discharge hoses.	❑	
10.13 Bilge				
10.13.1 ❑	Sewage, rodent excrement, food particles, putrescible matter or toxic substances found in the bilge.	Discharge the bilge water to a port-reception facility. Clean the bilge.		❑
		Check all parts of the sewage system for leakage and overflow.	❑	
		Perform deratting measures if rodent excrement is found.	❑	
10.13.2 ❑	Grey water regularly discharged into the bilge.	Collect grey water in the holding tank, lead it into the disinfection compartment of the sewage treatment plant or discharge it overboard according to local and international regulations.	❑	

Area 11 Ballast water

Introduction

Studies carried out in several countries have demonstrated that many species of bacteria, plants and animals are able to survive as "stowaways" in the ballast water and sediments carried by ships, even over long ocean voyages. Discharge of the ballast water and sediments in port waters can result in the establishment of harmful aquatic organisms and pathogenic agents that may pose a threat to human life, the environment and ecosystem balance.

Ships transporting large amounts of cargo (e.g. bulk ships or container ships) need to control their balance during cargo operations. Therefore, large amounts of ballast water are pumped into or out of the ship. For example, if a ship arrives in a port with empty cargo holds, the ship is "in ballast", which means that several hundred tonnes of ballast water are in the ballast-water tanks in order to stabilise the ship while crossing the ocean. During loading operations, the ballast water has to be pumped into the port basin to keep the ship stable.

International standards and recommendations

The IMO's International Convention for the Control and Management of Ships' Ballast Water and Sediments was adopted on 13 February 2004. The convention will enter into force 12 months after ratification by 30 states, representing 35% of the world's merchant-shipping tonnage (Article 18, Entry into force).

The specific requirements for ballast-water management are contained in Regulation B-3, Ballast water management for ships:

- Ships constructed before 2009 with a ballast-water capacity of between 1,500 and 5,000 m^3 must conduct ballast-water management that at least meet the ballast-water exchange standards or the ballast-water performance standards until 2014, after which it shall at least meet the ballast-water performance standard.
- Ships constructed before 2009 with a ballast water capacity of <1,500 m^3 or >5,000 m^3 must conduct ballast-water management that at least meets the ballast-water exchange standards or the ballast-water performance standards until 2016, after which time it shall at least meet the ballast-water performance standard.
- Ships constructed in or after 2009 with a ballast-water capacity of < 5,000 /m^3 must conduct ballast-water management that at least meets the ballast-water performance standard.
- Ships constructed in or after 2009, but before 2012, with a ballast-water capacity of ≥ 5,000 /m^3 shall conduct ballast-water management that at least meets the standard described in regulation D-1 or D-2 until 2016, and at least the ballast-water performance standard after 2016.
- Ships constructed in or after 2012 with a ballast-water capacity of ≥ 5,000 /m^3 shall conduct ballast-water management that at least meets the ballast-water performance standard.

Specific indicators for the ballast-water exchange standard and ballast-water performance standard are found in Annex D, Standards for ballast-water management, of the IMO International Convention for the Control and Management of Ships' Ballast Water and Sediments, and are described below.

Regulation D-1, Ballast water exchange standard:
Ships performing ballast-water exchange shall do so with an efficiency of 95% volumetric exchange of ballast water. For ships exchanging ballast water by the pumping-through method,

three times the volume of each ballast water tank must be pumped through to meet the described standard. Pumping through less than three times the volume may be accepted, provided the ship can demonstrate that at least 95% volumetric exchange is met.
Regulation D-2, Ballast water performance standard:
Ships conducting ballast-water management shall discharge:

- <10 viable organisms/m3 ≥ 50 μm in dimension;
- <10 viable organisms/ml <50 μm in dimension and ≥10 μm in dimension.

Also, discharge of the indicator microbes shall not exceed the specified concentrations.
The indicator microbes, as a human health standard, include, but are not limited to:

- toxicogenic Vibrio cholerae (O1 and O139) with <1 colony forming unit (cfu)/100 ml or <1 cfu/g (wet weight) of zooplankton samples;
- Escherichia coli <250 cfu/100 ml; and
- intestinal enterococci <100 cfu/100 ml.

Ballast-water management systems must be approved by the administration and accord with the IMO guidelines (Regulation D-3, Approval requirements for ballast water management systems). These include systems that make use of chemicals or biocides; make use of organisms or biological mechanisms; or alter the chemical or physical characteristics of the ballast water.

IMO guidelines for uniform implementation of the ballast water management convention:
Guidelines for sediments reception facilities (G1)
Guidelines for ballast water sampling (G2)
Guidelines for ballast water management equivalent compliance (G3)
Guidelines for ballast water management and development of ballast water management plans (G4)
Guidelines for ballast water reception facilities (G5)
Guidelines for ballast water exchange (G6)
Guidelines for risk assessment under Regulation A-4 (G7)
Guidelines for approval of ballast water management systems (G8)
Procedure for approval of BWM systems that make use of active substances (G9)
Guidelines for approval and oversight of prototype ballast water treatment technology programmes (G10)
Guidelines for ballast water exchange design and construction standards (G11)
Guidelines on design and construction to facilitate sediment control on ships (G12)
Guidelines for additional measures regarding ballast water management including emergency situations (G13)
Guidelines on designation of areas for ballast water exchange (G14)

Main risks

The problem of invasive species is largely due to the expanded trade and traffic volume over the past few decades. The effects in many areas of the world have been devastating. Quantitative data show that the rate of bio-invasions is increasing at an alarming rate, sometimes exponentially, and new areas are being invaded all the time. Seaborne trade continues to increase, and the problem may not yet have reached its peak.

Specific examples include the introduction of the European zebra mussel (Dreissena polymorpha) into the Great Lakes between Canada and the United States, resulting in billions of dollars being spent on pollution control and cleaning the underwater structures and water pipes; and the introduction of the American comb jelly (Mnemiopsis leidyi) to the Black and Azov seas, causing the near extinction of anchovy and sprat fisheries.

There are also public health risks; some cholera epidemics appear to be directly associated with ballast water in South America, the Gulf of Mexico and other areas.

Sampling

The standards of the IMO Guidelines MEPC 58/23 Annex 3, Draft guidelines for ballast water sampling (G2), can be used as a major reference for sampling ballast water, if necessary, for the assessment of existing public health risks.

The objectives of these guidelines are to provide the States Parties authorities, including Port State Control officers, with practical and technical guidance on ballast-water sampling and analysis for the purpose of determining whether the ship has complied with the ballast water management convention according to article 9, Inspection of ships, and with Regulations D-1 or D-2.

For this purpose, samples should be taken from the discharge line, as near to the point of discharge as practicable, during ballast-water discharge, whenever possible. In cases where the ballast system design does not enable sampling from the discharge line, other sampling arrangements may be necessary. Sampling via maintenance openings, sounding pipes or air pipes is not the preferred approach for assessing compliance with Regulation D-2.

Any sampling protocol for compliance testing under the ballast water management convention should observe the following principles to help ensure consistency of approach between parties and to provide certainty to the shipping industry:

- The sampling protocol should be in line with these guidelines.
- The sampling protocol should result in samples that are representative of the whole discharge of ballast water from any single tank or any combination of tanks being discharged.
- The sampling protocol should consider the potential for a suspended sediment load in the discharge to affect sample results.
- The sampling protocol should provide for samples to be taken at appropriate discharge points.
- The quantity and quality of samples taken should be sufficient to demonstrate whether the ballast water being discharged meets the relevant standard.
- Sampling should be undertaken in a safe and practical manner.
- Samples should be a manageable size.
- Samples should be taken, sealed and stored to ensure that they can be used to test for compliance with the convention.
- Samples should be fully analysed within the test method's holding-time limits by an accredited laboratory.
- Samples should be transported, handled and stored with consideration of the chain of custody.

Before testing for compliance with Regulation D-2, it is recommended that, as a first step, a representative sample of ballast-water discharge be taken to establish whether a ship is potentially compliant or non-compliant. Such a test could help the State Party identify immediate mitigation measures, within its existing powers, to avoid any additional impact from a possible non-compliant ballast-water discharge from the ship.

In emergency or epidemic situations, port states may use alternative sampling methods that may need to be introduced at short notice. Ships entering ports under their jurisdiction need to be told about these sampling methods. In such situations, they may not necessarily notify WHO, but such notification could be beneficial for other parties.

Document review

- Constructional drawings of ballast-water reporting system.
- IMO's ballast water reporting form.
- International safety management manual.
- Maintenance instructions for ballast-water treatment plant.

References

International conventions

IMO (2007), International Convention for the Control and Management of Ships' Ballast Water and Sediments.

Resolution MEPC. 152(55) *Guidelines for sediments reception facilities* (G1). London, IMO, 2005.

Resolution MEPC.123(53) *Guidelines for ballast water management equivalent compliance* (G3). London, IMO, 2005.

Resolution MEPC.124(53) *Guidelines for ballast water exchange* (G6). London, IMO, 2005.

Resolution MEPC.127(53) *Guidelines for ballast water management and development of ballast water management plans* (G4). London, IMO, 2005.

Resolution MEPC.140(54) *Guidelines for approval and oversight of prototype ballast water treatment technology programmes* (G10). London, IMO, 2006.

Resolution MEPC.149(55) *Guidelines for ballast water exchange design and construction standards* (G11). London, IMO, 2006.

Resolution MEPC.150(55) *Guidelines on design and construction to facilitate sediment control on ships* (G12). London, IMO, 2006.

Resolution MEPC.151(55) *Guidelines on designation of areas for ballast water exchange* (G14). London, IMO, 2006.

Resolution MEPC.153(55) *Guidelines for ballast water reception facilities* (G5). London, IMO, 2006.

Resolution MEPC.161(56) *Guidelines for additional measures regarding ballast water management including emergency situations* (G13). London, IMO, 2007.

Resolution MEPC.162(56) *Guidelines for risk assessment under Regulation A-4 of the BWM Convention* (G7). London, IMO, 2007.

Resolution MEPC.169(57) *Procedure for approval of ballast water management systems that make use of active substances*. London, IMO, 2008.

Resolution MEPC.173(58) *Guidelines for ballast water sampling* (G2). London, IMO, 2008.

Resolution MEPC.174(58) *Guidelines for approval of ballast water management systems* (G8). London, IMO, 2008.

Resolution MEPC.175(58) *Information reporting on type approved ballast water management systems*. London, IMO, 2008.

Resolution MEPC.188(60) *Installation of ballast water management systems on new ships in accordance with the application dates contained in the Ballast Water Management Convention (BWM Convention)*. London, IMO, 2010.

Scientific literature

McCarthy SA, Khambaty FM (1994). International dissemination of epidemic *Vibrio cholerae* by cargo ship ballast and other non-potable waters. *Applications in Environmental Microbiology*, 60:2597–2601.

Code of Areas	Inspection results: evidence found, sample results, documents reviewed	Control measures and corrective actions	Required	Recommended
11.1 Management				
11.1.1 ❏	IMO ballast-water record book not available.	Provide a complete ballast-water record book.	❏	
11.1.2 ❏	Ballast-water management plan (BWMP) not available.	Develop the BWMP according to IMO guidelines.		❏
		Implement all procedures as defined in the BWMP.		❏
11.1.3 ❏	Ballast-water treatment plants installed but no technical information available.	Provide technical information on the ballast-water treatment for next inspection.		❏
11.2 Ballast-water exchange and treatment				
11.2.1 ❏	No ballast-water exchange in the open sea has been performed, no onboard treatment system is available or no onboard treatment is performed.	Close all discharge valves immediately.	❏	
		Notify the proper authority (e.g. harbour police or Port State Control).	❏	
		Discharge, if necessary, under the supervision of the proper authority.	❏	
11.2.2 ❏	Salinity test indicates that water has not been exchanged in the open sea.	Inform the proper authority to collect samples for assessing the risk of harmful aquatic organisms and pathogens in the water.	❏	
11.2.3 ❏	Treatment plant not approved by IMO.	Close discharge lines and valves.		❏
		Discharge, if necessary, under supervision of the proper authority.		❏
		Notify proper authority (e.g. the harbour police or Port State Control).		❏
11.3 Discharge				
11.3.1 ❏	Untreated or unexchanged ballast water has been or is discharged into the port basin, the river or another protected area.	Stop discharge operation immediately and notify proper authority (e.g. the harbour police or Port State Control).	❏	

Area 12 Cargo holds

Introduction

Factors contributing to onboard public health risks include the design, construction, management and operation of the cargo holds. Some public health risks can be carried from one country to another via contaminated or infested cargo loaded into holds, onboard contamination or vector infestation of cargo, and inadequate or insufficient onboard preventive and control measures.

Holds should be empty for inspection. According to IMO Resolution A.864(20) and the Manual on loading and unloading of solid bulk cargoes for terminal representatives (BLU Code), 2008 edition, special precautions should be taken before entering enclosed spaces aboard ships. There could be a risk of an unsafe atmosphere in ships' holds, particularly where the cargo has been fumigated in passage, and/or has oxygen-depleting characteristics, or flammable or toxic vapours. Hold inspections should be carried out as soon as unloading of a hold is completed and it is safe to enter.

International standards and recommendations

IMO Recommendations for entering enclosed spaces aboard ships [resolution A.864(20)]

3 Assessment

3.2 The procedures to be followed for testing the atmosphere in the space and for entry should be decided on the basis of the preliminary assessment. These will depend on whether the risk assessment shows that:

1. there is minimal risk to the health or life of personnel entering the space;
2. there is no immediate risk to health or life, but a risk could arise during the course of work in the space; and
3. a risk to health or life is identified.

Where the preliminary assessment indicates minimal risk to health or life, or potential for risk to arise during the course of work in the space, the precautions described in 4, 5, 6 and 7 should be followed as appropriate.

Where the preliminary assessment identifies risk to life or health, and if entry is necessary, the additional precautions specified in section 8 should also be followed.

9.5 Fumigation

When a ship is fumigated, the detailed recommendations contained in the recommended use of pesticides in ships should be followed. Spaces adjacent to fumigated spaced should be treated as if fumigated.

IMO I267:2008, Manual on loading and unloading of solid bulk cargoes for terminal representatives (BLU Code), 2008 edition

This manual comprises regulations to prevent pollution from household garbage and other solid waste. The annex defines the different types of waste that are to be regarded as garbage, the distance from land where they are allowed to be discharged and in what way.

International Convention for the Prevention of Pollution from Ships 1973 (modified by the protocol of 1978 relating thereto [MARPOL 73/78])

According to this convention, the following types of waste are regarded as garbage from cargo holds: dunnage, broken pallets, lashings, ropes and covers.

IMO SOLAS XII/6.5.1—Protection of cargo holds from loading/discharge equipment and SOLAS XII/6.5.3—Failure of cargo hold structural members and panels

Main risks

Main public health risks on board include the design, construction, management and operation of the cargo holds. Some public health risks can be carried from one country to another via contaminated or infested cargo loaded into holds, onboard contamination or vector infestation of cargo, and inadequate or insufficient onboard preventive and control measures.

Document review

- Management plans.
- Operational procedures.
- Confined-space entry procedures and records.
- Lock-out and tag-out procedures.
- Material Safety Data Sheets.
- Construction drawings (including drainage).
- Ventilation system drawings.
- Vector control records.
- Cleaning schedule.

References

International conventions

IMO, International Convention for the Safety of Life at Sea (SOLAS), Regulation XII/6.5.1. London, IMO, 1974.

IMO, I267:2008 *Manual on loading and unloading of solid bulk cargoes for terminal representatives.* London, IMO, 2008.

IMO, *Recommendations for entering enclosed spaces aboard ships.* London, IMO, 1997.

Code of Areas	Inspection results: evidence found, sample results, documents reviewed	Control measures and corrective actions	Required	Recommended
12.1 Operational procedures				
12.1.1 ❑	Absent or ineffective operational procedures for controlling public health risks according to the characteristics and forms of cargo on board.	Lay out operational procedures to proactively control public health risks to staff, travellers and communities that could be affected by ships and cargoes arriving in port.		❑
		Develop appropriate management plans.		❑
12.2 Construction, design and layout				
12.2.1 ❑	Construction, design and layout predispose to probable failure of public health risk controls.	Perform measures to correct construction, design and layout to make them suitable for intended purpose.		❑
12.2.2 ❑	Ingress of contaminated materials, liquids, gases, foreign material or vectors found.	Perform disinfection, disinsection or deratting if contamination is evident.	❑	
		Separate cargo suspected or showing signs of contamination or deterioration.		❑
12.3 Cleaning and maintenance				
12.3.1 ❑	Construction materials and design do not facilitate cleaning, and/or construction design aids harbouring of vectors.	Correct design deficiencies, and reconstruct using materials that facilitate cleaning and decontamination.		❑
		Discard or isolate contaminated items.	❑	
12.3.2 ❑	Evidence of vectors and/or reservoirs found.	Perform immediate disinfection, disinsection or deratting.	❑	
12.4 Equipment for controlling environmental conditions				
12.4.1 ❑	Absent, inadequate or ineffective equipment necessary for controlling environmental conditions according to type of cargo.	Correct equipment deficiencies and write procedures to ensure effective operational practices.		❑
12.5 Drainage				
12.5.1 ❑	Drains not independent of each other and of all other drainage systems.	Correct deficiencies and ensure that drains are independent of each other and of all other drainage systems.	❑	
12.5.2 ❑	Drains connected to a drain that receives human sewage or medical waste.	Separate drain from any drainage system carrying human or medical waste.	❑	
12.5.3 ❑	Drain lines do not discharge to open drain wells.	Drain lines to discharge to open drain wells with an air gap.	❑	
12.6 Ventilation				
12.6.1 ❑	Service outlet of the cold-air or hot-air system (or both) serves more than one compartment.	Install separate ducts for ventilation, air-conditioning and heating systems.		❑
		Separate the service outlets for each compartment.		❑
12.6.2 ❑	Evidence of vectors and/or reservoirs found.	Vector-proof the ducts extending from the weather deck directly to the cargo holds, engine room and boiler rooms, with no horizontal extensions, at either end.		❑
		Perform immediate disinfection, disinsection or deratting.	❑	

Area 13 Other systems and areas

Introduction

There are other systems and areas that also present health concerns. Vectors pose a major risk to the health of passengers and crew members. On board, mosquitoes, rats, mice, cockroaches, flies, lice and rat fleas are all capable of transmitting disease. Also, rodents are well established at port areas and are considered vectors for many diseases, such as plague, murine typhus, salmonellosis, trichinosis, leptospirosis and rat-bite fever. Monitoring and control of vectors and reservoirs is necessary to maintain health on ships.

Standing water caused by heavy rainfall or overflow can act as breeding sites for mosquitoes. This can then increase the potential for exposure to vector-borne diseases such as dengue fever, malaria and West Nile fever.

Washing machines and laundry facilities are indispensible to ships, according to C92 Accommodation of Crews Convention (Revised) 1949. This document contains detailed minimum standards for the location, construction, arrangement and equipment of such facilities, including sanitary facilities.

Drainage from laundries is categorized as grey water, under Annex V of the International Convention for the Prevention of Pollution from Ships 1973 (modified by the protocol of 1978 relating thereto [MARPOL 73/78]). Under the previous Garbage Pollution Prevention Regulations, liquefied galley wastes were not considered to be waste, and there were therefore no restrictions on their discharge as long as they did not contain a pollutant as described in the MARPOL 73/78 regulations. Although grey water generally poses less harm than, for example, black water, grey water can on occasion contain some harmful constituents, such as detergent residues and chlorine from bleach in laundry discharges.

International standards and recommendations

International Health Regulations (IHR) (2005)

Article 24, Conveyance operators:

States Parties shall take practicable measures consistent with these regulations to ensure that conveyance operators:

(a) comply with the health measures recommended by WHO and adopted by the State Party;

(b) inform travellers of the health measures recommended by WHO and adopted by the State Party for application on board; and

(c) permanently keep conveyances for which they are responsible free from sources of infection or contamination, including vectors and reservoirs. The application of measures to control sources of infection or contamination may be required if evidence is found.

Annex 1B, To provide as far as practicable a programme and trained personnel for the control of vectors and reservoirs in and near points of entry.

Annex 5, Specific measures for vector-borne diseases.

ILO Maritime Labour Convention 2006

Regulation 3.1, Accommodation and recreational facilities.

Standard A3.1, Accommodation and recreational facilities.

13. Appropriately situated and furnished laundry facilities shall be available.

Regulation 3.1, Accommodation and recreational facilities
Guideline B3.1.7, Sanitary accommodation
4. The laundry facilities provided for seafarers' use should include:
(a) washing machines;
(b) drying machines or adequately heated and ventilated drying rooms; and
(c) irons and ironing boards or their equivalent.

ILO C92 Accommodation of Crews Convention (Revised) 1949
Article 13:
12. In all ships, facilities for washing and drying clothes shall be provided on a scale appropriate to the size of the crew and the normal duration of the voyage.
13. The facilities for washing clothes shall include suitable sinks, which may be installed in wash rooms, if separate laundry accommodation is not reasonably practicable, with an adequate supply of cold fresh water and hot fresh water or means of heating water.
14. The facilities for drying clothes shall be provided in a compartment separate from sleeping rooms and mess rooms, adequately ventilated and heated and equipped with lines or other fittings for hanging clothes.

Main risks

Many diseases are transmitted to humans via vectors such as rats, mosquitoes, mice, cockroaches, flies, lice and rat fleas. If not properly controlled, these vectors could board ships, and then breed and be carried overseas. This would represent serious health risks to the crew and passengers. Moreover, persons and vectors on board can, in turn, spread disease to ports in other countries.

For example, standing water on board the ship or on its lifeboats provide a habitat for mosquitoes to lay eggs. Adult mosquitoes will emerge from the standing water, and these adults will lay additional eggs during their life-cycle. If standing water persists for long periods, or is replenished by repeated heavy rain or overflow, increased mosquito production may continue for several weeks or months. If not effectively controlled, mosquitoes could be carried by ship and spread infectious disease via international travel.

Other risks include exposure to blood or other potentially infectious materials through contaminated objects that were improperly handled during housekeeping; for instance, the housekeeping staff contact contaminated laundry because they were not wearing appropriate personal protective equipment (PPE). Finally, the presence of hazardous chemicals used in the laundry process, dust from clothes and powder detergents, and poor ventilation of the workplace all present health risks.

Document review

Integrated vector management plan.

References

International conventions

ILO, Maritime Labour Convention 2006.

Scientific literature

Anselmo M et al. (1996). Port malaria caused by *Plasmodium falciparum*: a case report. *Le Infezioni in Medicina*, 4:45–47.
Delmont J et al. (1994). Harbour-acquired *Plasmodium falciparum* malaria. *Lancet*, 344:330–331.

Delmont J et al. (1995). Apropos of 2 cases of severe malaria contracted in the port of Marseille. *Bulletin de la Societe de Pathologie Exotique*, 88:170–173.

Draganescu N et al. (1977). Epidemic outbreak caused by West Nile virus in the crew of a Romanian cargo ship passing the Suez Canal and the Red Sea on route to Yokohama. *Virologie*, 28:259–262.

Fijan S, Sostar-Turk S, Cencic A (2005). Implementing hygiene monitoring systems in hospital laundries in order to reduce microbial contamination of hospital textiles. *Journal of Hospital Infection*, 61(1):30–38.

Peleman R et al. (2000). Indigenous malaria in a suburb of Ghent, Belgium. *Journal of Travel Medicine*, 7:48–49.

Raju N, Poljak I, Troselj-Vukic B (2000). Malaria, a travel health problem in the maritime community. *Journal of Travel Medicine*, 7:309–313.

Rubin L, Nunberg D, Rishpon S (2005). Malaria in a seaport worker in Haifa. *Journal of Travel Medicine*, 12:350–352.

Schultz MG et al. (1967). An outbreak of malaria on shipboard. *American Journal of Tropical Medicine and Hygiene*, 16:576–579.

Shoda M et al. (2001). Malaria infections in crews of Japanese ships. *International Maritime Health*, 52:9–18.

Code of Areas	Inspection results: evidence found, sample results, documents reviewed	Control measures and corrective actions	Required	Recommended
	13.1 Overall vector management system			
13.1.1 ❑	No rat-proof guard.	Place rodent-proof guard to prevent rodents from boarding ships via the mooring lines.	❑	
13.1.2 ❑	No integrated vector management plan.	Develop an integrated vector management plan.	❑	
13.1.3 ❑	No vector control inspection records and logs available (including pesticide application).	Conduct routine surveillance on vectors and reservoirs; for example, deploy and check rodent traps and other devices.	❑	
		Develop vector control inspection records and logs, including pesticide application logs.	❑	
	13.2 Standing water			
13.2.1 ❑	Evidence of standing water in different areas of the ship's open spaces (e.g. lifeboat covers, bilges, scuppers, awnings, gutters, air-treatment plants) that can hold insect larvae. Evidence of depressions or culverts that can collect standing water.	Implement operational procedures to control public health risks to crew and passengers, and communities that could be affected by ships and cargoes arriving in ports.		❑
13.2.2 ❑	Evidence of live vectors or their larvae in standing water inside lifeboats.	Eliminate standing water and apply vector control measures.	❑	
	13.3 Construction of washing machines and laundry			
13.3.1 ❑	Construction materials and design make cleaning difficult.	Redesign and reconstruct materials so that they facilitate cleaning and decontamination.		❑
13.3.2 ❑	Improper installation of soil and waste drainage systems.	Ensure soil and waste drainage system is of adequate dimensions, and constructed to minimize the risk of obstruction and facilitate cleaning.		❑
13.3.3 ❑	Doors and windows not conducive for proper ventilation.	Correct door and window design.		❑
13.3.4 ❑	Floors not conducive for cleaning.	Reconstruct floors using durable material that is easily cleaned and impervious to dampness, and properly drains.		❑
	13.4 Housekeeping cleaning and maintenance			
13.4.1 ❑	Evidence of housekeeping crew members cleaning cabins of ill passengers or crew while not wearing PPE.	Ensure that the housekeeping crew members take precautions, including using disposable PPE that is changed after cleaning each ill person's cabin.	❑	
13.4.2 ❑	Insufficient supply of detergents.	Supply housekeeping and laundry with sufficient quantity of washing powder or similar products.		❑
13.4.3 ❑	Evidence of leaks, overflow or cross-connection in the drainage system.	Maintain drainage system so that it does not leak or back up.	❑	
13.4.4 ❑	Evidence of housekeeping crew members using the same wiping cloth to clean cabins of ill passengers or crew, as well as cabins of well passengers or crew; or to first clean cabins of ill passengers and crew.	Change wiping cloth following cleaning of cabins of ill passengers and crew.	❑	
		Clean cabins of well passengers and crew before ill passengers and crew.	❑	

Annex 1 Model Ship Sanitation Control Exemption Certificate/ Ship Sanitation Control Certificate, Annex 3 of International Health Regulations (2005)

Annex 3

Model Ship Sanitation Control Exemption Certificate/Ship Sanitation Control Certificate

Port of ………. Date: …………..

This Certificate records the inspection and 1) exemption from control or 2) control measures applied

Name of ship or inland navigation vessel ……....................Flag …...................... Registration/IMO No. …………......

At the time of inspection the holds were unladen/laden with ...... tonnes of .... cargo

Name and address of inspecting officer …………………..

Ship Sanitation Control Exemption Certificate

Areas, [systems, and services] inspected	Evidence found[1]	Sample results[2]	Documents reviewed
Galley			Medical log
Pantry			Ship's log
Stores			Other
Hold(s)/cargo			
Quarters:			
- crew			
- officers			
- passengers			
- deck			
Potable water			
Sewage			
Ballast tanks			
Solid and medical waste			
Standing water			
Engine room			
Medical facilities			
Other areas specified - see attached			
Note areas not applicable, by marking N/A.			

No evidence found. Ship/vessel is exempted from control measures.

Ship Sanitation Control Certificate

Control measures applied	Re-inspection date	Comments regarding conditions found

Control measures indicated were applied on the date below.

Name and designation of issuing officer …………………………………........... Signature and seal …………………………… Date ………….......

[1] (a) Evidence of infection or contamination, including: vectors in all stages of growth; animal reservoirs for vectors; rodents or other species that could carry human disease, microbiological, chemical and other risks to human health; signs of inadequate sanitary measures. (b) Information concerning any human cases (to be included in the Maritime Declaration of Health).

[2] Results from samples taken on board. Analysis to be provided to ship's master by most expedient means and, if re-inspection is required, to the next appropriate port of call coinciding with the re-inspection date specified in this certificate.

Sanitation Control Exemption Certificates and Sanitation Control Certificates are valid for a maximum of six months, but the validity period may be extended by one month if inspection cannot be carried out at the port and there is no evidence of infection or contamination.

Attachment to model Ship Sanitation Control Exemption Certificate/Ship Sanitation Control Certificate

Areas/facilities/systems inspected	Evidence found	Sample results	Documents reviewed	Control measures applied	Re-inspection date	Comments regarding conditions found
Food						
• Source						
• Storage						
• Preparation						
• Service						
Water						
• Source						
• Storage						
• Distribution						
Waste						
• Holding						
• Treatment						
• Disposal						
Swimming pools/spas						
• Equipment						
• Operation						
Medical facilities						
• Equipment and medical devices						
• Operation						
• Medicines						
Other areas inspected						

Indicate when the areas listed are not applicable by marking N/A.

Annex 2 Algorithm for issuance of ship sanitation certificates, handling of re-inspections and affected conveyances

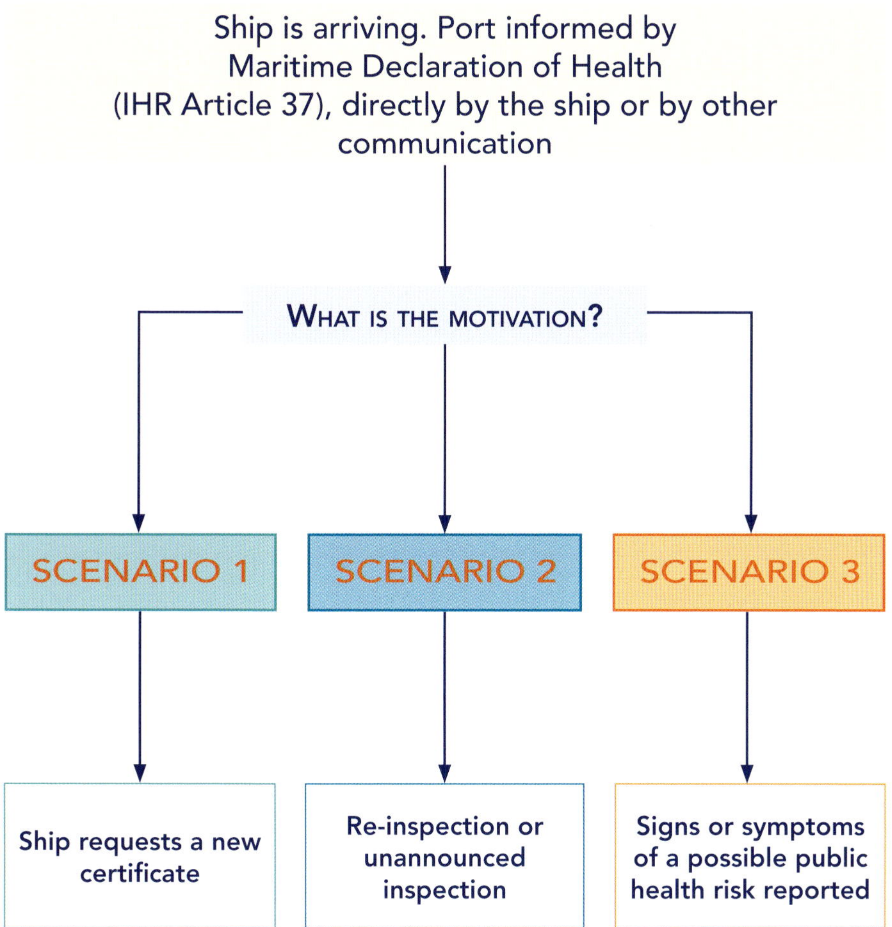

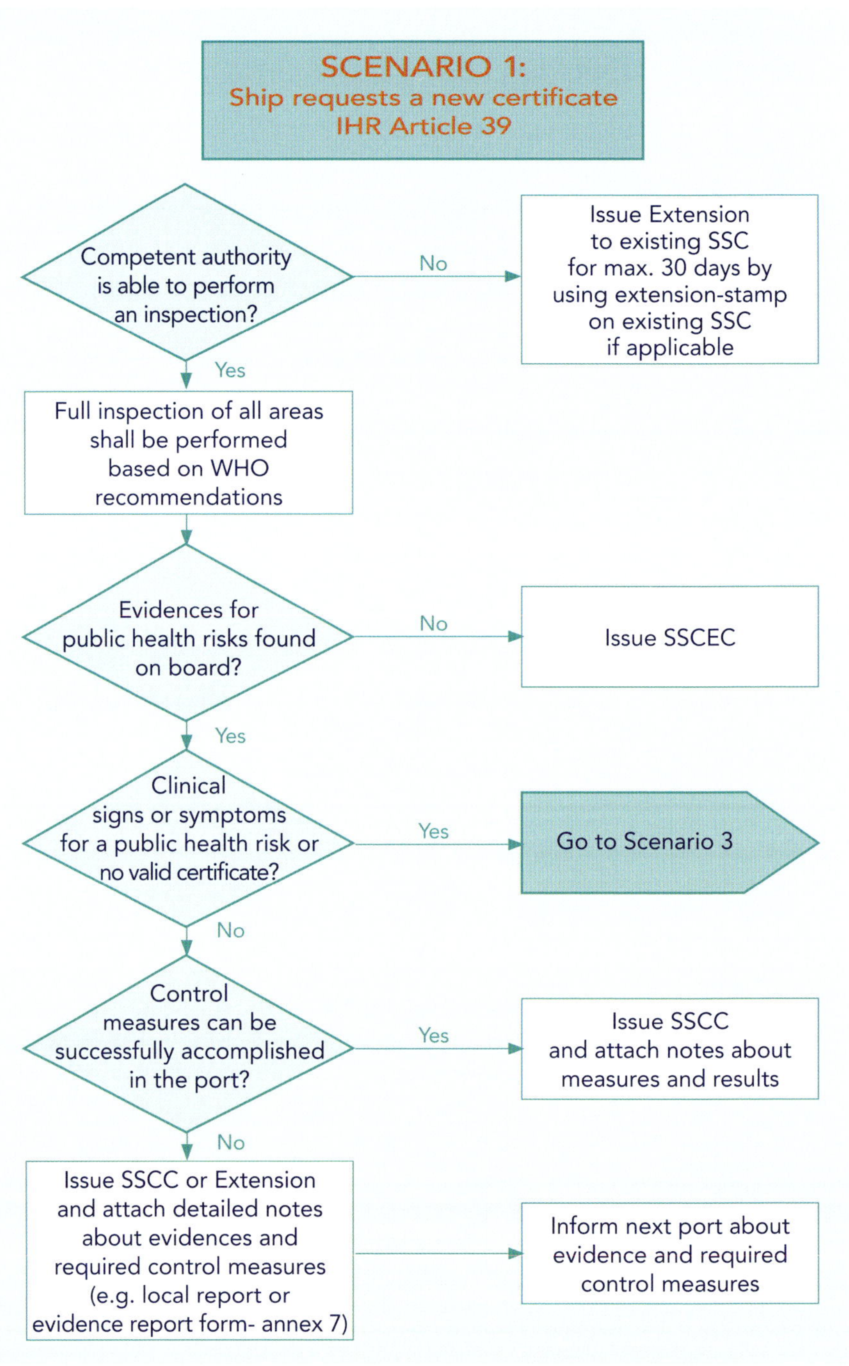
SCENARIO 1:
Ship requests a new certificate
IHR Article 39
Competent authority is able to perform an inspection?
No
Issue Extension to existing SSC for max. 30 days by using extension-stamp on existing SSC if applicable
Yes
Full inspection of all areas shall be performed based on WHO recommendations
Evidences for public health risks found on board?
No
Issue SSCEC
Yes
Clinical signs or symptoms for a public health risk or no valid certificate?
Yes
Go to Scenario 3
No
Control measures can be successfully accomplished in the port?
Yes
Issue SSCC and attach notes about measures and results
No
Issue SSCC or Extension and attach detailed notes about evidences and required control measures (e.g. local report or evidence report form- annex 7)
Inform next port about evidence and required control measures

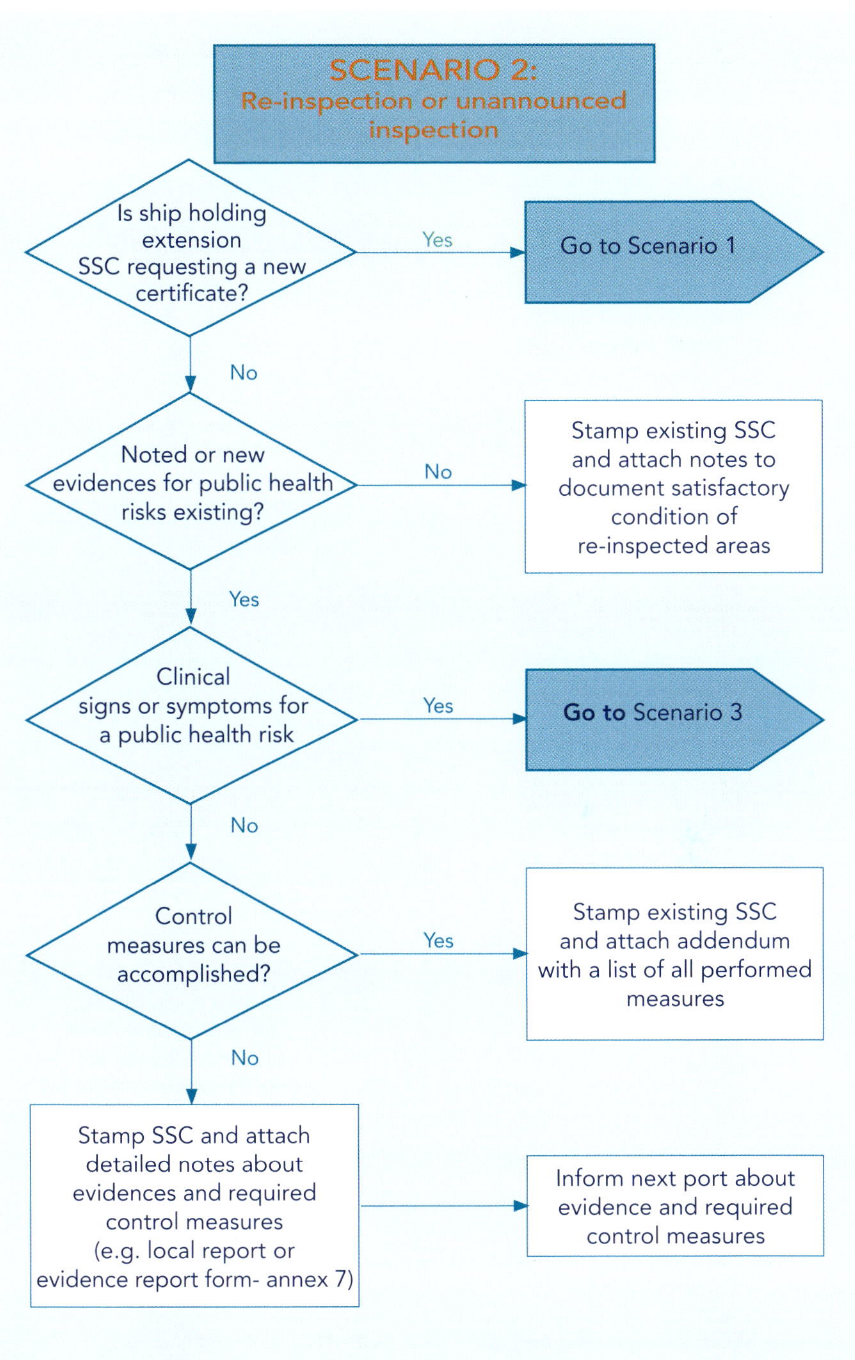

SCENARIO 2:
Re-inspection or unannounced inspection
Is ship holding extension SSC requesting a new certificate?
Yes
Go to Scenario 1
No
Noted or new evidences for public health risks existing?
No
Stamp existing SSC and attach notes to document satisfactory condition of re-inspected areas
Yes
Clinical signs or symptoms for a public health risk
Yes
Go to Scenario 3
No
Control measures can be accomplished?
Yes
Stamp existing SSC and attach addendum with a list of all performed measures
No
Stamp SSC and attach detailed notes about evidences and required control measures (e.g. local report or evidence report form- annex 7)
Inform next port about evidence and required control measures

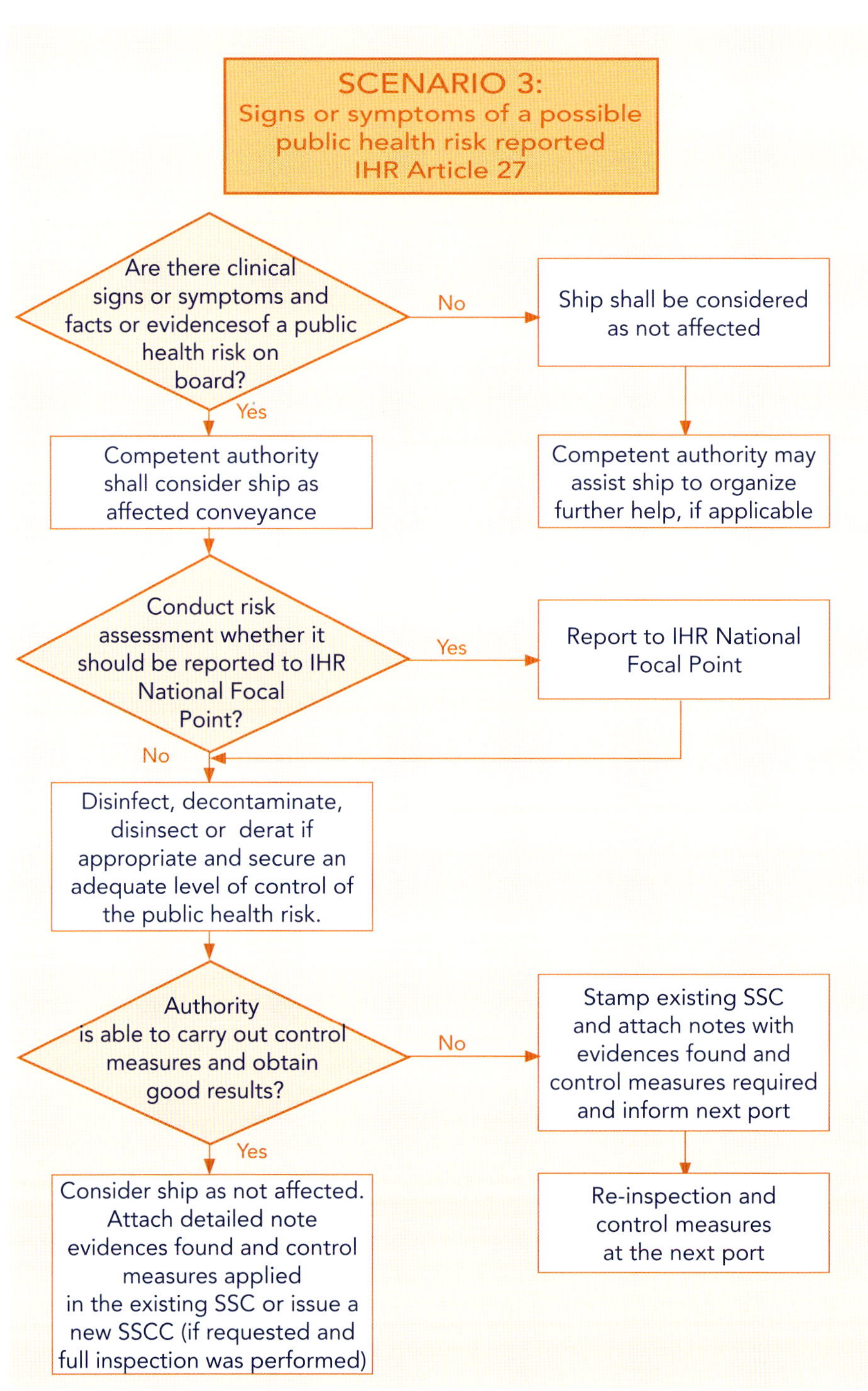
SCENARIO 3:
Signs or symptoms of a possible public health risk reported
IHR Article 27
Are there clinical signs or symptoms and facts or evidencesof a public health risk on board?
No
Ship shall be considered as not affected
Competent authority may assist ship to organize further help, if applicable
Yes
Competent authority shall consider ship as affected conveyance
Conduct risk assessment whether it should be reported to IHR National Focal Point?
Yes
Report to IHR National Focal Point
No
Disinfect, decontaminate, disinsect or derat if appropriate and secure an adequate level of control of the public health risk.
Authority is able to carry out control measures and obtain good results?
No
Stamp existing SSC and attach notes with evidences found and control measures required and inform next port
Re-inspection and control measures at the next port
Yes
Consider ship as not affected. Attach detailed note evidences found and control measures applied in the existing SSC or issue a new SSCC (if requested and full inspection was performed)

Annex 3 Sequence of inspection areas

This annex proposes a sequence of inspection areas to be implemented in the ship's procedures manual. It follows the rationale that clean areas should be inspected first, followed by technical areas. This sequence enables the inspectors to avoid cross-contamination.

In some critical areas like galleys, inspectors should demonstrate good hygiene practice by wearing clean disposable clothing (e.g. aprons, gloves, hair covering).

Table A3.1
Proposed sequence of inspection areas

Region	Area	Rationale
Inside accommodation	Quarters	Start at the top of the accommodation
	Galley, pantry and service areas	Potentially more clean than pantry
	Pantry	Potentially more clean than stores
	Stores	Close to galley and pantry
	Child-care facilities	Usually more dirty than food areas
	Medical facilities	After food areas to avoid cross-contamination
	Swimming pools, spas and saunas	Sometimes inside, sometimes on deck
	Other areas and systems	Washing, laundry usually cleaner than waste
	Waste (solid and medical)	Most dirty area in accommodation, sometimes on deck
Inside engine room	Engine room	Overview
	Potable water	Most parts in the engine room
	Sewage	Most parts in the engine room
	Ballast tanks	Access from the engine room, through the pipe duct, from open deck or cargo compartments
Outside	Cargo holds	Outside
	Standing water	On deck

Annex 4 Personal protective equipment for inspectors and crew

The following table is a list of personal protective equipment (PPE) that should be available to ship staff and inspectors. For inspectors, the table indicates if they should wear the PPE during a normal inspection, or during an outbreak investigation.

Table A4.1
Personal protective equipment available to crew and inspectors

Personal protective equipment item	Inspector during normal inspection	Inspector in case of outbreak investigation
Safety helmet	X	X
Hair net	X	
Safety goggles or face shield	X	
Ear protection	X	
Face mask		X (FFP3 standard)
Working gloves (e.g. leather gloves)	X	
Rubber examination gloves	X	X
Kitchen apron	X	
Watertight apron	X	
Disposable overalls		X
Hand disinfection liquid	X	X
Signal vest or signal jacket	X	
Inflatable life vest	X	
Safety shoes with non-slip and anti-sparkle soles	X	
Insect repellent in some areas	X	

Annex 5 Technical equipment useful to perform a ship inspection

The following table lists technical equipment that could be available to ship inspectors to help them inspect ships. It also indicates if the inspectors should use the equipment during a normal inspection, or if it is more suitable for use during an outbreak investigation.

Table A5.1 Technical equipment useful for ship inspectors

Technical equipment	Inspector during normal inspection	Inspector in case of outbreak investigation	Function
Flashlight (ideally explosion proof)	X	X	Explosion-proof design can be useful in some areas.
Calibrated food-probe thermometer (contact or infrared)	X	X	To measure food temperatures. Infrared thermometers are useful to avoid direct contact.
Vermin indicator spray	X	X	To cast out cockroaches from sealed spaces.
White cloth	X	X	To find vermin (e.g. fleas).
Double-faced adhesive tape	X	X	To detect crawling insects.
Seals and stamps	X	X	To authenticate certificates.
Pens, clipboard and notepad	X	X	To facilitate paperwork.
Dictionary	X	X	To facilitate communication between ship operator and inspector.
Screwdriver kit	X	X	To open devices for inspection where necessary.
First-aid kit	X	X	For personal safety.
Laptop and mobile printer	X	X	Can be useful to print certificates or define line lists directly from a database or other software.
Foldable ruler or measuring tape	X		To measure the size of air gaps, other dimensions and so on.
Smoke pen or other smoke-generating devices	X	X	To test exhaust hoods and ventilation systems.
Ultraviolet (UV) flashlight	X	X	To detect urine contamination by human and rodents.

Technical equipment	Inspector during normal inspection	Inspector in case of outbreak investigation	Function
Water-testing kit that includes: • pH meter • thermometer • conductivity sensor • chlorine-testing kit • hardness-testing kit • turbidity-testing kit	X	X	To estimate risk of possible contamination of the on-board potable water system and to be able to survey disinfection measures.
Water sampling kit that includes: • gripper and hex wrench • gas burner • ethanol spray (70%) • disposable paper towels.	X	X	To take water samples at a high enough quality to be analysed.
Sterile glass bottles containing sodium thiosulfate	X	X	For microbiological water analysis.
Protein-detecting swab	X		To check appropriate cleaning of surfaces (e.g. in galley).
Sample containers for: • stool and urine samples • blood samples • swabs • food samples		X	To collect samples of different possible sources of contamination (e.g. water, food, human, surfaces, equipment).
Camera (ideally digital)	X	X	To capture evidence.

Due to safety reasons and to facilitate entering onto the ship, the equipment on this list should be stored in a backpack or other suitable portable container(s).

Annex 6 Model documents for ship inspection

This section provides a list of documents that need to be available during a ship inspection. This list should be sent by the issuing authorities before the ship arrives at the port to facilitate preparation for the ship inspection.

Table A6.1
List of model certificates and documents required for ship sanitation inspections

Category	Name	Note
A IHR-related documents	Maritime Declaration of Health	IHR Annex 8
	Ship sanitation control certificate	IHR Annex 3
	Ship sanitation control exemption certificate	
	Extension of the ship sanitation certificate	
	International Certificate of Vaccination or Prophylaxis	IHR Annex 6
B Other documents, as listed on the IMO Convention on Facilitation of International Maritime Traffic 1965 (as amended, 2006 edition), might be requested for assessment of public health risks	General declaration	http://www.imo.org/Facilitation/mainframe.asp?topic_id=396
	Cargo declaration	http://www.imo.org/Facilitation/mainframe.asp?topic_id=396
	Ships Stores Declaration	http://www.imo.org/Facilitation/mainframe.asp?topic_id=396
	Crew list	http://www.imo.org/Facilitation/mainframe.asp?topic_id=396
	Passenger list	http://www.imo.org/Facilitation/mainframe.asp?topic_id=396
	Dangerous goods manifest	http://www.imo.org/Facilitation/mainframe.asp?topic_id=396
	International Sewage Pollution Prevention Certificate	
	Garbage management plan	Every ship that is ≥ 400 gross tonnage and every ship that is certified to carry >15 persons shall carry a garbage management plan, which the crew must follow: http://www.imo.org/Facilitation/mainframe.asp?topic_id=396
	Garbage record book	Same as above http://www.imo.org/Facilitation/mainframe.asp?topic_id=396
	Crews effects declaration	http://www.imo.org/Facilitation/mainframe.asp?topic_id=396
	Document required under the Universal Postal Convention (for mail)	In the absence of such a document, the postal objects (number and weight) must be shown on the cargo declaration
	Others	
C Other management plans concerning onboard hygiene	Water safety plan (or water management plan)	
	Potable water analysis report	
	Waste management plan	
	Management plan for food safety (including food temperature record)	

Continuation of the annex 6

Category	Name	Note
C Other management plans concerning onboard hygiene	Management plan for vector control	
	Medical log	
	IMO ballast water reporting form	
	Ballast Water Management Certificate	http://www.imo.org/about/conventions/listofconventions/pages/international-convention-for-the-control-and-management-of-ships'-ballast-water-and-sediments-(bwm).aspx
	Ballast-water record book	http://www.imo.org/about/conventions/listofconventions/pages/international-convention-for-the-control-and-management-of-ships'-ballast-water-and-sediments-(bwm).aspx
	Other	

Annex 7 Evidence Report Form

Table A7.1 is a sample of an Evidence Report Form. This form lists the evidence found, samples and documents reviewed, and control measures or corrective actions to be performed after a ship inspection, and supports the ship sanitation certificate (SSC). When attached to the SSC, each page of this attachment needs to be signed, stamped and dated by the issuing authority. If this form is used as an attachment to a pre-existing SSC, this attachment must be noted in the SSC (e.g. by using a stamp).

Table A7.1
Sample Evidence Report Form

Evidence Report Form This form supports the ship sanitation certificate (SSC), and provides a list of evidence found and control measures to be performed. When attached to the SSC, each page of this attachment needs to be signed, stamped and dated by the competent authority. If this document is used as an attachment to a pre-existing SSC, this attachment must be noted in the SSC (e.g. by using a stamp).	
Ship's name and IMO no. or registration:	Name and signature of responsible onboard ship officer :
Name of issuing authority:	Actual inspection date (dd/mm/yyyy):
Date of referred SSC (dd/mm/yyyy):	SSC issued in the port of:

Indicate areas that <u>have not</u> been inspected:			
❑ Quarters	❑ Galley, pantry, service area	❑ Stores	❑ Child-care facilities
❑ Medical care facilities	❑ Swimming pools/spas	❑ Solid and medical waste	❑ Engine room
❑ Potable water	❑ Sewage	❑ Ballast water	❑ Cargo holds
❑ Other (e.g. laundry and washing machine)			
Detected heath events on board		❑ yes	❑ No

Evidence code	Evidence found (brief description according to WHO checklist; draw a line under each item of evidence to ensure items are clearly separated)	Measure to be applied	Required	Recommended	Measure successfully performed (stamp and signature of re-inspecting authority)

Name of issuing inspector:	Signature of issuing inspector:	Stamp of issuing authority:	Page.. of..

IMO, International Maritime Organization; SSC, ship sanitation certificate; WHO, World Health Organization.

Annex 8 Instructions for completing the Evidence Report Form

1. At the top of the form, the following information should be documented:
- ship name and International Marine Organization (IMO) or registration number
- name and signature of the responsible ship officer
- name of issuing authority
- date of actual inspection
- date of the ship sanitation certificate (SSC) that the Evidence Report Form refers to
- port where the SSC was issued.

2. The areas that have not been inspected are indicated by the checklist.

3. Further information in the form:

1st column, Evidence code	Code of evidence as described in the checklist
2nd column, Evidence found	Brief description of the evidence found based on the checklist
3rd column, Measures to be applied	Description of the control measure(s) that should be applied (use simple words as in checklist)
4th column, Required	Note an "X" here if the measure should be a "requirement"
5th column, Recommended	Note an "X" here if the measure should be a "recommendation"
6th column, Measures successfully performed	This column is reserved for the re-inspecting authority. Only successfully performed measures should be stamped and signed by the re-inspecting issuing authority. If no stamp or signature is applied, the success of the measure should be verified in a new inspection. To differentiate between each piece of evidence and the corresponding measures, horizontal lines should be drawn to separate different evidence.

At the bottom of each page of the form, the name and signature of the inspecting officer, the stamp of the issuing authority, the page number and any comments should be noted.

References and resources

CAC (Codex Alimentarius Commission) (2003)., *Recommended international code of practice—general principles of food hygiene* (incorporates Hazard Analysis and Critical Control point [HACCP] system and guidelines for its application), CAC/RCP 1-1969, Rev. 4-2003. Rome, CAC.

CAC (Codex Alimentarius Commission) (2004).*Code of hygienic practice for milk and milk products*, CAC/RCP 57. Rome, CAC.

CDC (Centers for Disease Control and Prevention). Vessel Sanitation Program. National Center for Environmental Health. Atlanta, CDC
(http://www.cdc.gov/nceh/vsp).

Helsinki Commission (1990). HELCOM Recommendation 11/10: *Guidelines for capacity calculation of sewage system on board passenger ships*. Helsinki, Helsinki Commission.

ILO (International Labour Organization) (2006). Maritime Labour Convention 2006. Geneva, ILO.

IMO (International Maritime Organization) (1978). Annex IV: Prevention of pollution by sewage from ships and corresponding resolutions MEPC.2(VI) and MEPC.115(51). In: International Convention for the Prevention of Pollution from Ships 1973 (amended 1978) (MARPOL 73/78). London, IMO.

IMO (International Maritime Organization) (1978). Annex V: Prevention of pollution by garbage from ships. In: International Convention for the Prevention of Pollution from Ships 1973 (amended 1978) (MARPOL 73/78). London, IMO.

IMO (International Maritime Organization) (1982). *Medical first aid guide for use in accidents involving dangerous goods* (MFAG). London, IMO.

IMO (International Maritime Organization) (1995). Emergency, occupational safety, medical care and survival functions. In: International Convention on Standards, Certification and Watchkeeping for Seafarers 1978 (amended 1995). London, IMO.

IMO (International Maritime Organization) (1996). Resolution MEPC.70(38) *Guidelines for the development of garbage management plans*. London, IMO.

IMO (International Maritime Organization) (1997). Resolution A.868(20) *Guidelines for the control and management of ships' ballast water to minimize the transfer of harmful aquatic organisms and pathogens*. London, IMO.

IMO (International Maritime Organization) (1997). Resolution MEPC.76(40) *Standard specification for shipboard incinerators*. London, IMO.

IMO (International Maritime Organization) (2000). *Guidelines for ensuring the adequacy of port waste reception facilities*. London, IMO.

IMO (International Maritime Organization) (2004). International Convention for the Control and Management of Ships' Ballast Water and Sediments. London, IMO.

IMO (International Maritime Organization) (2006). Annex 5: Certificates and documents required to be carried on board ships. In: Convention on Facilitation of International Maritime Traffic 1965 (2006 ed.). London, IMO.

IMO (International Maritime Organization) (2008). *Guidelines for the uniform implementation of the BWM convention*. London, IMO.

ISO (International Organization of Standardization) (2002). ISO 15748-1:2002 *Ships and marine technology—potable water supply on ships and marine structures*. Geneva, ISO.

ISO (International Organization of Standardization) (2006). ISO 19458:2006 *Water quality—sampling for microbiological analysis*. Geneva, ISO.

ISO (International Organization of Standardization) (2008). ISO 14726:2008 *Ships and marine technology—identification colours for the content of piping systems*. Geneva, ISO.

ISO (International Organization of Standardization), IEC (International Electrotechnical Commission) (2005). ISO/IEC 17025:2005 *General requirements for the competence of testing and calibration laboratories*. Geneva, ISO/IEC.

WHO (World Health Organization) (1999). *Safe management of wastes from health-care activities*. Geneva, WHO.

WHO (World Health Organization) (2007). *International medical guide for ships*, 3rd ed. Geneva, WHO.

WHO (World Health Organization) (2008). *Guidelines for drinking-water quality*. Geneva, WHO.

WHO (World Health Organization) (2011). *Guide to ship sanitation*, 3rd ed. Geneva, WHO.